The United States Patent System

Legal and Economic Conflicts in American Patent History

FLOYD L. VAUGHAN

The United States Patent System

Legal and Economic Conflicts in American Patent History

GREENWOOD PRESS, PUBLISHERS
WESTPORT, CONNECTICUT

The Library of Congress has catalogued this publication as follows:

Library of Congress Cataloging in Publication Data

Vaughan, Floyd Lamar, 1891-
The United States patent system.

Bibliography: p.
1. Patent laws and legistation--United States.
2. Patents--United States. I. Title.
[KE3114.V38 1972] 608'.7'73 72-6846
ISBN 0-8371-6499-0

Originally published in 1956 by the University of Oklahoma Press, Norman

Reprinted with the permission of the University of Oklahoma Press

Reprinted by Greenwood Press, Inc.

First Greenwood reprinting 1972
Second Greenwood reprinting 1977

Library of Congress catalog card number 72-6846
ISBN 0-8371-6499-0

Printed in the United States of America

PREFACE

THIS BOOK GOES much beyond the scope of *Economics of Our Patent System,* the first of my ventures into the complex field of patent control. Patents still provide the prospect of remuneration and stimulate inventive activity, even though they seldom enable the inventor to realize financial returns. Moreover, they are still used in violation of antitrust legislation, and a high proportion of them are either invalid or suppressed. The records in support of these conclusions, far more voluminous now than thirty years ago, still include court decisions, government reports and hearings, and individual studies. In this book, as in these records, more space is necessarily devoted to abuses little known than to alleged benefits generally admitted.

It was possible for me to return to this field of research because of assistance from several institutions—a grant from the Social Science Research Council in 1948, a reduction in teaching hours by the University of Oklahoma in 1948–49, and a Fulbright grant for research at the London School of Economics in 1949–50. The economists and lawyers who criticized parts or all of the manuscript were very helpful in pointing out errors and suggesting improvements, and I wish to acknowledge my indebtedness to them.

In the footnotes of this book, I have used abbreviated titles for those Congressional hearings and reports which are repeatedly cited. A list of the abbreviations used, with a complete identification of each, will be found on page 319.

FLOYD L. VAUGHAN

CONTENTS

The United States Patent System

Legal and Economic Conflicts in American Patent History

CHAPTER I

Background of the Problem

THE HISTORY OF MAN is largely a record of invention. The wheeled vehicle, the compass, and gunpowder facilitated mobility, exploration, and conquest. The printing press became the medium for the education of the masses. Power-driven machinery ushered in the Industrial Revolution. More recent inventions in transportation and communication have contributed to man's conquering of space and time. New weapons of destruction threaten the existence of the human race as advances in medicine still extend the span of life.

The new technology—power from steam, gasoline, and electricity; invention and development of machines; the development of new processes—has resulted in large-scale production of goods and services, both old and new. Still bigger and newer things are in the making, and others unheard of presumably are still to come. Organized research, along with the inquiry of the lone inventor, continues today the long tradition of man's inventiveness—an inventiveness that yet provides the equipment, in the largest sense of the word, for man's material progress and at the same time heightens the pressure and tensions of contemporary life, and the resulting search for security, with the production of the increasingly amazing articles—both constructive and destructive—which man's twentieth-century know-how makes possible.

Causes of Invention

But what is the cause of invention? Does the inventor, like the painter or poet, merely respond to the creative urge? May the esteem of others provide a sufficient incentive for his efforts? Does the mere repetition of the same task suggest a labor-saving device? Is philosophical speculation the mental attitude most conducive to invention? Is the kind and extent of invention dependent largely upon the state of science and the technical arts? What influence does the scarcity of labor have on invention and does the quick and widespread adoption of certain inventions, such as the radio and television serve as an inspiration for the inventor? Would inventions occur in the absence of patents? These questions, involving various aspects of man and his environment, deserve answers, however tentative and qualified they may be.

Instinct of Contrivance. According to Thorstein Veblen, "The instinct of workmanship . . . occupies the interest with practical expedients, ways and means, devices and contrivances of efficiency and economy, proficiency, creative work and technological mastery of facts."[1] F. W. Taussig states that "the biographies of inventors give abundant illustrations of the state of inward happiness which comes from the exercise of the contriving bent."[2] According to Francis A. Walker, the power to invent, which inheres to so remarkable a degree in our people, was created altogether irrespective of, and long antecedent to, a system of patent legislation.[3] Inventors themselves have testified to the influence of the instinct of contrivance upon invention. Thomas A. Edison once stated that the inventor will continue to invent "as along as he commands a dollar. This is a natural peculiarity of the inventive mind."[4] The instinct of contrivance is undoubtedly a stronger force in men like Edison than in others who merely improve basic inventions.

Esteem of Others. Other factors, such as the desire for public esteem, are conducive to invention. Inventors have often complained of the lack of appreciation of their efforts. "For inventors have ever been depreciated in their day; even at the present time, despite the known facts as to what inventions and inventors have done for every one of us, the inventor as an inventor is lightly regarded. So are his inventions until they have ceased to be regarded as inventions, and have been accepted as constituent parts of the machine of civilization. By that time the inventor has often been forgotten."[5]

A government may offer titles, official acclaim, and other kinds of public recognition in stimulating inventions. Napoleon employed

1 *The Instinct of Workmanship*, 33.

2 *Inventors and Money-Makers*, 15.

3 *Discussions in Economics and Statistics*, II, 182.

4 Nolan Hearings of 1919, 173. (For complete identification of this and other citations of government hearings and reports, see p. 319.

5 Bradley A. Fiske, *Invention, The Master-Key to Progress*, 19.

these appeals as well as monetary rewards to acquire inventions that would help meet the crisis brought on by a naval blockade; for example, he acquired an invention that would make possible the deriving of sugar from beets. Appreciation by the public undoubtedly means much to men like Edison and Louis Pasteur, though they may never admit it.

Division of Labor. Adam Smith observed "that the invention of all those machines by which labor is so much facilitated and abridged seems to have been originally owing to the division of labor. Men are much more likely to discover easier and readier methods of attaining any object when the whole attention of their minds is directed towards that single object than when it is dissipated among a great variety of things. But in consequence of the division of labor, the whole of every man's attention comes naturally to be directed towards some one very simple object. It is naturally to be expected, therefore, that some one or other of those who are employed in each particular branch of labor should soon find out easier and readier methods of performing their own particular work, wherever the nature of it admits of such improvement."[6] John Hobson states, "Nearly all the great textile inventors were practical men, most of them operatives immersed in the details of their craft, brought face to face continually with some definite difficulty to be overcome, some particular economy desirable to make."[7] Those who make and operate machines have the opportunity to see what goes on—an experience that brings forth ideas for improvement. It is for this reason that the research laboratory is needed less for machinery than for chemicals.

There are many exceptions to the alleged importance of the division of labor in fostering inventions; for example, Richard Arkwright was a barber, S. F. B. Morse a portrait painter, Eli Whitney a teacher, and Wilbur and Orville Wright were bicycle repairmen.[8] It should be added that such incongruity is found also in other fields of endeavor; for example, David Ricardo, a banker, made keen observations about agriculture and on the basis of his observations built his theory about the distribution of wealth.

Intellectual Speculation. Adam Smith, however, realized that di-

6 *The Wealth of Nations,* Bk. I, Chap. I, Par. 8.
7 *The Evolution of Modern Capitalism,* 80.
8 Comp. L. Hearings, 244–45.

vision of labor, as exemplified by manual labor, was not the only cause of invention. Men possessed of intellectual curiosity, who represent one aspect of the division of labor, may bring forth inventions of the most fundamental sort. As he said, "All the improvements in machinery, however, have by no means been the inventions of those who had occasion to use the machines. Many improvements have been made by the ingenuity of the makers of the machines, when to make them became the business of a peculiar trade; and some by that of those who are called philosophers or men of speculation, whose trade it is not to do anything, but to observe everything; and who, upon that account, are often capable of combining together the powers of the most distant and dissimilar objects."[9]

Professors Joseph Henry of Princeton and S. M. Babcock of Wisconsin discovered the principles of the telegraph and the cream separator respectively while in the pursuit of academic investigations, and both of them gave their inventions to the world. The patent system evidently played no part in stimulating their intellectual inquiries. Henry said, "I never myself attempted to reduce the principles to practice, or to apply any of my discoveries to processes in the arts. My whole attention, exclusive of my duties to the College, was devoted to original scientific investigations, and I left to others what I considered, in a scientific view, of subordinate importance—the application of my discoveries to useful purposes in the arts. Besides this I partook of the feeling common to men of science, which disinclines them to secure to themselves the advantages of their discoveries by a patent."[10] Thomas Jefferson refused to take out patents on his inventions. Benjamin Franklin likewise had a mind of the inquiring bent and "never deigned to patent any of his inventions."[11] Edmund Cartwright—"a man leading a life of thought"—was one of the great textile inventors.[12]

Inventions arising from intellectual inquiry have been explained in some instances as accidents or the results of divine inspiration. P. G. Hubert tells a story concerning Charles Goodyear in which chance plays an important part. "Standing before a stove in a store at Woburn, Mass., he was explaining to some acquaintances the properties

[9] *The Wealth of Nations,* Bk. I, Chap. I, Par. 9.

[10] Holland Thompson, *The Age of Invention,* 200.

[11] *Ibid.,* 9.

[12] Hobson, *The Evolution of Modern Capitalism,* 80.

of a piece of sulphur-cured india-rubber which he held in his hand. They listened to him good-naturedly, but with evident incredulity, when suddenly he dropped the rubber on the stove, which was red hot. His old clothes would have melted instantly from contact with such heat; but, to his surprise, this piece underwent no change. In amazement he examined it, and found that while it had charred or shrivelled like leather, it had not softened at all. The bystanders attached no importance to this phenomenon, but to him it was a revelation."[13] Some of Franklin's discoveries relating to electricity are partly attributable to accident. However, his conceptions, like those of other inventors, came into being as the result of pertinent lines of inquiry. An Englishman named William Henry Perkin, while working in a laboratory to make a synthetic cleaner, made the classic discovery of the mauve color in one of his vats,[14] from which has developed the coal tar dyes. The oil-flotation method for recovering small particles of copper ore was discovered by accident. "That was a discovery; yet, if the fellow hadn't been working in that field, he wouldn't have made the discovery."[15] A more recent illustration of the same idea is the discovery of penicillin by Dr. Alexander Fleming of Great Britain. While he was working on some bacterial cultures, the cover of one of the dishes in which they were being grown was left off. A few days later he noticed—as many another scientist had observed—that a spot of mold had formed. A mold spore had fallen and had begun to grow. The bacteria around the spot of mold had apparently disappeared while those some distance from it had continued to increase. Fleming, his curiosity aroused, grew more of the mold. He found that when diluted even five hundred times, the mold was two or three times as effective as pure carbolic acid. Most important, it was the first agent he had found which in destroying bacteria did not also injure white blood corpuscles. Another recent example is the discovery of a preventive for travel sickness incidental to the search for a remedy for hay fever. A patient who was taking the drug reported no improvement, but remarked incidentally that she was not nauseated by the train trips from her home town as in the past. Further experiments confirmed the effectiveness of this drug for travel sickness. Scientific inquiry, not mere accident, explained the result.

[13] P. G. Hubert, *Men of Achievement—Inventors,* 167–68.
[14] Pooling Hearings, Part 1, 750.
[15] TNEC Hearings, Part 2, 353.

Inventors and others, presumably with a strong religious bent, have attributed to God their inventions as well as other creative works. Goodyear wrote, "The creature imagines he is executing some plan of his own, while he is simply an instrument in the hands of his Maker for executing the divine purposes of beneficence to the race." These words "are a fitting preface for the story of the years that followed, which is a tale of endurance and persistent activity under sufferings and disappointments such as are scarcely paralleled even in the pages of invention, darkened as they often are by poverty and defeat."[16] Goodyear, like many men who have contributed much, died a pauper. One writer quotes David Livingstone to the effect that "the existence of various implements which are in use among the Africans and other partially civilized races, points to the communication of an instruction which must have proceeded at some time or another from a superhuman being."[17]

Accumulation of Knowledge. Intellectual speculation is facilitated by our accumulation of past knowledge. We have, as W. F. Ogburn once phrased it, a "technical heritage." There is, historically speaking, a correlation between the advancement of science and the arts, and the amount and kind of invention. The dissemination of this learning by general education enlarges, of course, its effect on invention. According to Veblen, "Any new technological departure necessarily takes its rise in the workmanlike endeavors of given individuals, but it can do so only by force of their familiarity with the body of knowledge which the group already has in hand."[18] The inventors of the steam engine, for example, received inestimable aid from physics. "It was by the knowledge which they gleaned regarding the properties of steam and air and water and iron, regarding the laws of motion and heat and work and force and weight and mass, that the inventors' experiments were guided."[19]

A considerable part of the knowledge which furnishes a foundation and stimulation to invention is invention itself. Nothing invents like invention. No invention can be considered absolute. Wireless communication was developed up to the point of the radio invention before M. G. Marconi became the acknowledged "inventor" of it. The

[16] Thompson, *Age of Invention,* 165.
[17] William I. Thomas, *Source Book for Social Origins,* 431.
[18] *The Instinct of Workmanship,* 104.
[19] Fiske, *Invention, The Master-Key to Progress,* 156.

airplane utilizes the internal combustion engine and other prior inventions and also the machinery and tools—the result of invention—necessary in putting the various parts together. Inventions thus resemble the stones composing a pyramid in that each of them rests upon the others. Anthropologists declare that "every art that is used to minister to our comfort in this present vastly complex civilization has been brought forward step by step, beginning with the simple needs of cruder times. . . . The development of invention is like the unfolding of the human mind."[20] And Hobson says, "No one of the inventions which were greatest in their effect, the jenny, the water-frame, the mule, the power-loom, was in the main attributable to the effort or ability of a single man; each represented in its successful shape the addition of many successive increments of discovery; in most cases the successful invention was the slightly superior survivor of many similar attempts."[21]

However, the possession of and strict adherence to accumulated knowledge may be a handicap. Much of the so-called duplication of effort in acquiring knowledge already available is not waste. In no line of important endeavor can one simply begin where others have left off; a build-up of experience is necessary. Finding the way through a jungle is more apt to lead to alternate paths and even new discoveries if the trees are not already blazed. Moreover, ignorance about the impossible may be helpful in certain instances. Simon Newcomb, dean of American science at the turn of the century, wrote in 1903: "May not our mechanicians be ultimately forced to admit that aerial flight is one of that great class of problems with which man can never hope to cope and give up all attempts to grapple with it?"[22] The Wright brothers either failed to read or did not heed this statement. Along this same line someone has observed, "According to recognized aeronautical tests the bumble bee cannot fly because of the shape and weight of his body in relation to the total wing area. But the bumble bee doesn't know this, so goes ahead and flies anyway." Charles F. Kettering has said, "If you want a thing done, don't give it to the man familiar with the art, who knows it cannot be done; give it to

20 Walter Hough, *Synoptic Series of Objects in the United States National Museum Illustrating the History of Inventions*, 2–3.

21 *The Evolution of Modern Capitalism*, 79.

22 NRC Report, 4.

someone who does not know it cannot be done—and he will do it."[23] Of course one should not infer that in general the fetters of knowledge are more restrictive than the shackles of ignorance.

Economic Conditions. The process of conceiving and contriving inventions requires a fund of goods to live on and work with while the process is under way. Individuals who live from hand to mouth are not likely to invent, even though other conditions are favorable. As W. E. Hearn once stated, "If an example be sought of the different rates at which inventive talent grows where capital is abundant and where it is scarce, it is sufficient to compare the England of the nineteenth with the England of the eighteenth century."[24] Evidence of this need for capital prior to invention is the fact that, according to the president of E. I. Du Pont de Nemours & Co., research and development and manufacturing facilities costing $27,000,000 and a waiting period of ten years preceded the commercial production of nylon.[25]

The actual or potential demand for an invention may be the controlling factor. Writing about 1790, M. G. J. de Crèvecœur, the famous French observer of America, said, "The farmer and the artisan have more to do than they can perform; scarcity of men makes labor very dear; to supply the want of labor and time the American is forced to invent, to think out new ways of augmenting his efficiency."[26] An important part of public demand for new inventions is purchasing power and willingness to buy. These factors in demand are encouraged by installment plans and sales campaigns so that a high-priced invention like the electric refrigerator can be made and sold on a large scale soon after its technical development and thus permit the manufacturer and even the inventor to make money from it. In the United States people can and will buy, and this willingness and ability is important in encouraging the development of new goods and services for the consumer. Even when there is demand and available support for an invention, however, a new product does not always come about. Existing equipment may prohibit the manufacture of the new; patents on new equipment may be acquired and suppressed merely to prevent

[23] Comp. L. Hearings, 371.

[24] *Plutology*, 268.

[25] MP Hearings, Part 2–A, 546.

[26] Pooling Hearings, Part 4, 3593.

the scrapping of the old. Also, new products, inventions, and the like, may have to conform to inventions already in existence; for example, any technical improvement of trains must conform to two rails 4 feet 8½ inches apart and to tunnels of fixed sizes.

There is plenty of evidence to show that even in the face of demand inventions are sometimes not forthcoming. "It is often said that 'necessity is the mother of invention,' but the history of the constructive work of this country appears to disprove the aphorism. At no time in the life of our people has the need of labor-saving contrivances been so great as during the first hundred years after the settlement of the country; yet, as we have noted, this was, as far as mechanical contrivances were concerned, a peculiarly barren period."[27] According to another writer, "The invention of the sewing machine was one rich influence on subsequent progress; and all the story connected with it is interesting in many ways. But the most wonderful fact connected with the invention is that it was not made before."[28] It is interesting to note Veblen's reversal of the relation that exists between necessity and invention: "And here and now, as always and everywhere, invention is the mother of necessity."[29] In other words, once an invention is devised, it becomes necessary to use it in a competitive regime.

Similar conditions foster the conception of a particular invention by different inventors at the same time. "Many minds of a similar character are thus set upon the same train of thought and investigation. It is therefore natural that these similar inquiries should in at least some cases lead to the same conclusion."[30]

Economic Reward. The most prevalent explanation of the cause of invention is the desire for economic reward by means of a patent. According to G. Friedrich List, "The granting of patent privileges offers a prize to inventive minds."[31]

But history shows little correlation between the grant of monopoly rights to inventors and the emergence of great inventions. The inventiveness of man, a characteristic which sets him apart from other animals, manifested itself many centuries before the granting of pat-

[27] N. S. Shaler (ed.), *The United States of America,* II, 135.
[28] Fiske, *Invention, The Master-Key to Progress,* 236.
[29] *The Instinct of Workmanship,* 314.
[30] Hearn, *Plutology,* 195–96.
[31] *The National System of Political Economy,* 246.

ents. People of ancient times had such important inventions as the wheeled vehicle, the needle, the loom, the potter's wheel, metal working, the plow, the rotary drill, the screw press, the pulley, the steelyard, fire-producing means, glass, the lathe, and the water wheel. By the close of the fifteenth century, other inventions included the mariner's compass, paper and the printing press, the mechanical clock, iron casting, the spinning wheel, fulling stocks, gunpowder, spectacles, and the windmill. In most instances it is not known by whom, when, or where they were made. There is no evidence of patent grants anywhere which might have fostered them. Would a patent system have brought forth these inventions more quickly? Would it have led to more and better results?

Although the granting of monopoly rights for new manufactures, whether invented or imported, became well established during Queen Elizabeth's reign, there was no appreciable change in technological ideas or methods in England for over two hundred years. Why did the Industrial Revolution not occur much earlier in response to monopoly grants for inventors? Because conditions were right for it in the eighteenth, not the seventeenth or sixteenth, century. The crude beginnings of industry offered a field in which the instinct of contrivance could find expression. Once invention got under way, as in textile machinery, different inventors were a challenge to each other, and division of labor became feasible. Intellectual inquiry had a rapidly growing fund of knowledge to feed upon. Free man, capital reserves, and a wide market supplanted serfdom, a hand-to-mouth existence, and local self-sufficiency. All these things as well as a patent system provided a favorable environment for invention. The same inquiry and the same analysis apply to America, except that these forces and the subsequent great industrial awakening here came later.

The presidents of three automobile companies—General Motors, Ford, and the former Packard company—along with other business executives, have testified that their research would continue even though they could not take out patents on their inventions. According to these executives, research is the effort to solve a problem rather than to obtain a patent, and the main purpose in taking out patents is to keep from being excluded by others rather than to exclude them. However, various persons who contend that patents are not essential to the research of large corporations, and therefore to their improvements, some of which may be patented, admit the importance of patents to the small company and the lone inventor. The patent counsel of the Ford Motor Company, for example, testified that "we don't get most of our progress from what we call patent-minded men as we do from men who are looking forward to promoting progress," and yet conceded "unquestionably there have been many contributions of value which have come from men who would not have done their work had they not had before them at some time the hope of the reward of a patent."[32] Any discussion of patents as a cause of invention must differentiate between hired and independent inventors.

History of Patents

An understanding of patents requires an inquiry into their history not only in the United States but in England, for the patent laws of this country have been greatly influenced by those of the mother country. A brief history of the origin and development of the patent system will therefore be presented.

In England. The promotion of trade and industry formed one of the early policies of England, manifested first by the granting of special privileges to the towns and to the merchant and craft guilds. These grants, in contrast to later patents, created a group privilege under control of the crown rather than a specific monopoly under the control of one person. Edward III instituted a policy of granting letters of protection to foreign workmen. In 1331, he granted such a letter to John Kempe, a weaver of woolen cloth, "to exercise and teach his trade in this country."[33] A statute of 1337 confirmed such a grant and gave further aid by prohibiting the export of wool and the import of foreign cloth.

These letters of protection to foreigners led to special grants to domestic workers, illustrated by monopolies on a new process for making salt and the introduction of the manufacture of white soap, in order to foster new industries. As has been indicated, the new industry might be the result of inventive ingenuity at home or the introduction of a manufacture from abroad. In addition, monopolistic privileges were given to the East India Company and other trading companies to promote foreign commerce.

32 TNEC Hearings, Part 2, 263, 265.

33 A. A. Gomme, *Patents of Invention,* 10.

There are several explanations for these patents of monopoly. During the sixteenth century, England, in comparison with France and other parts of Europe, lagged behind in economic development. The creation of new industries required, it seemed, a special stimulus in the guise of monopolistic privileges. In this respect patents served the purpose of a protective tariff. Perhaps the desire of Elizabeth I to strengthen the political power and prestige of the nation was another factor in her willingness to create monopolies that were national in scope and subservient to the crown. Furthermore, the sovereigns of the time, continually embarrassed by depleted exchequers, contrived many devices to replenish them, one of which consisted of granting exclusive monopolies in return for royalties, although in most instances the expense of protecting the patentee from infringement left little revenue. Another reason for these patent monopolies was the desire to reward favorites of the court, many of whom had performed valuable services. The presence of unscrupulous favorites and the need of revenue eventually led to an abuse of the patent system; the promotion of new industries gave way to a scheme of odious exploitation. For a small pittance most of the life necessities were farmed out as private monopolies. Further, some patents conferred the power of supervision over trade or industry, and others granted special licenses for the evasion of laws which restricted the import, export, or movement of certain commodities. In 1601, Queen Elizabeth I, because of threatened action by Parliament, summarily cancelled the most objectionable patents and allowed the courts of law to pass judgment upon the remainder. It should be added that this patent controversy was only a part of the struggle for power between the crown and Parliament, one phase of which was the clash between the crown and the common law courts.

The first opportunity of the court to exercise its new prerogative came in 1602, when a certain manufacturer, Darcy by name, brought suit against Allin for the infringement of his exclusive patent for importing, making, and selling playing cards. The following words, spoken by a lawyer during the proceedings and accepted by the judges in the case as a true statement of law, defined the conditions under which a right to a monopoly should be upheld: "When any man by his own charge and industry, or by his own wit or invention doth bring any new trade into the realm, or any engine tending to the furtherance of a trade that never was used before; and that for the good of the realm; that in such cases the king may grant to him a monopoly-patent for some reasonable time, until the subjects may learn the same, in consideration of the good that he doth bring by his invention to the commonwealth, otherwise not."[34]

Edward Coke, the judge of the case, described a monopoly so as to leave a loophole for patent grants: "A monopoly is an institution or allowance by the King—by his grant, commission, or otherwise, to any person or persons, bodies politic or corporate, of or for the sole buying, selling, making, working, or using of anything whereby any person or persons, bodies politic or corporate, are sought to be restrained of any freedom or liberty that they had before, or hindered in their lawful trade."[35] Thus, a patent grant for a few years, covering either an invention or a new industry or a new trade, does not restrain the people of any freedom or liberty that they had before, nor does it hinder them in their lawful activity. The American colonies, as will be noted later, incorporated the spirit of Coke's decision and definition into their own patent laws.

The successor to Elizabeth I, James I, disregarding the decision of the court, greatly exploited the patent system in an effort to secure pecuniary aid; so mischievous and oppressive became the practice that in 1623 Parliament wrung from him the Statute of Monopolies. This statute declared all monopolies contrary to the laws of England, but provided an exception: "That any declaration beforementioned shall not extend to any letters-patent or grants of privilege for the term of fourteen years, or under, hereafter to be made of the sole working or making of any manner of new manufacture within this realm, to the true and first inventor of such manufactures, which others, at the time of making such letters-patents and grant, shall not use, so as also they be not contrary to the law nor mischievous to the State by raising prices of commodities at home or hurt of trade, or generally inconvenient. . . ." The exception, be it noted, was "the sole working or making of any manner of new manufacture"; in other words, it was a new industry arising from either inventive ingenuity at home or introduction from abroad.

In the Colonies. The special monopolies granted by the original

[34] *Darcy* v. *Allin,* 11 Coke 84b (K. B. 1602).
[35] Coke, 3 Institute 181.

thirteen colonies were not limited to new inventions, but like the British laws were used also to encourage the introduction of new manufactures from other countries, or to assist a manufacturer to engage in a familiar industry, or to facilitate inspection. The early legislators, however, prescribed definite boundaries to these privileges. The colonial law of Massachusetts, for example, stipulated, "No monopolies shall be granted or allowed amongst us, but of such new Inventions that are profitable to the Countrie, and that for a short time."[36]

Massachusetts and the other colonies granted numerous monopolies, some of which will be briefly described. In 1641, the General Court of Massachusetts Bay issued a patent for a novel method of making salt. In 1646, the same court granted a monopoly on engines for sawmills, but the colony retained the power to restrain the exportation of the inventions and to moderate their prices.[37] A ten-year patent, in 1750, to a manufacturer of sperm candles and other whale products, required him to teach at least five apprentices during that period, two of whom should be nominated by the General Court.[38]

In 1691, Connecticut offered a monopoly in salt making to any person who would set up the first works. In 1717, it conditioned a monopoly, covering the making of molasses from corn stalks, upon the petitioners producing "as good molasses and as cheap as comes from the West Indies."[39] This colony was the most liberal in the number of its grants for the promotion of the useful arts.

New sources of trade were always encouraged in Rhode Island, and patent rights were readily granted to any who would introduce desirable branches of industry. An exclusive right for ten years was bestowed upon a Portuguese to manufacture soap, similar to that made in Castile, of which he knew the process. At the expiration of his patent, he was to reveal the secret. The disclosure of the invention, it should be noted, was to occur when the patent expired and not when it was granted.[40]

Plows were patented and in use in Virginia before the assembling

36 *Body of Liberties of Massachusetts,* 1641, Liberty 9.

37 *Records of the Colony of Massachusetts Bay,* May 6, 1646, III, 65.

38 Victor S. Clark, *History of Manufactures in the United States,* I, 50.

39 *Ibid.*

40 Samuel G. Arnold, *History of the State of Rhode Island,* II, 228.

of the First Congress under the Constitution of the United States.[41] Moreover, Virginia granted money to several inventors as a consideration for their inventions. One inventor, for example, addressed a petition to the Governor and Council "setting forth that he had at great trouble and expense invented a new compass and protractor, by which an angle may be measured both in surveying and platting with greater Accuracy than by any other instrument hitherto discovered and praying such Bounty as the Legislature may think he deserves and the said petition was read." The legislature allowed him thirty pounds as a consideration for his useful invention.[42]

It appears that South Carolina alone gave patent protection to new machinery during the last century of colonial history. In 1784 it enacted, according to Frank I. Schechter, "probably the first general patent law, as distinguished from an individual legislative grant of a patent," and it provided for compulsory licensing in case of suppression.[43]

After the Revolution, mechanical improvements received more public attention in all the states. John Fitch and James Rumsey were allowed patents for steamboats by several states, and, in 1787, the important improvements made by Oliver Evans in mill machinery were similarly protected. Mechanical inventions received relatively less legislative protection during the colonial period than later because tools rather than machinery were used for manufacturing and power was employed in but few industries.[44]

New York apparently was the only colony that used monopolies as subsidiary to the inspection laws, to assure standard products. This is probably the explanation of the tanning monopoly in New York City granted in 1676 by the British colonial governor Edmund Andros to two tanners, for it was given at the same time that laws were passed regulating the industry.[45]

The temporary and local effect of these special monopolies may have been at times considerable, but their permanent and general influence upon colonial manufacturers might almost be disregarded. Their common use, in New England especially, to encourage new

[41] *Proceedings and Addresses of the Patent Centennial Celebration, 1891,* 65.

[42] *Journal,* House of Burgesses, Dec. 6, and Dec. 15, 1764.

[43] Pooling Hearings, Part 4, 3570–71.

[44] Clark, *History of Manufactures in the United States,* 49.

[45] *Ibid.,* 52.

industries appears to have been due to an acquired habit of legislation rather than to successful experience. After 1750 there was less of this method of promoting industries. Exclusive privileges continued to be granted to inventors, and these later laws mark the transition from the monopolies of an earlier date to the adoption of a patent system in the Constitution and subsequent statutes.

In the United States. Each colony had granted patents independently of the others. The Articles of Confederation of 1781 permitted the continuation of this policy. The arrangement led to conflict, confusion, and uncertainty, and therefore suggested the desirability of a national patent system. Incidentally, an interesting phase of this was the steamboat patent litigation which culminated in *Gibbons* v. *Ogden*[46] in 1824—a landmark in the interpretation of the commerce clause of the Constitution. Because of such conflict, the Constitution empowers Congress "to promote the progress of science and the useful arts by securing for limited times to authors and inventors, the exclusive rights to their respective writings and discoveries."[47] According to a recent government report,

> The background, the balanced construction, and the usage current then and later, indicate that the constitutional provision is really two provisions merged into one. The purpose of the first provision is to promote the progress of science by securing for limited times to authors the exclusive right to their writings, the word "science" in this connection having the meaning of knowledge in general, which is one of its meanings today. The other provision is that Congress has the power to promote the progress of useful arts by securing for limited times to inventors the exclusive right to their discoveries. The first patent law and all patent laws up to a much later period were entitled "Acts to promote the progress of useful arts."[48]

Under authority of the constitutional provision, Congress passed its first patent law in 1790. An examining board, composed of the secretary of state, the secretary of war, and the attorney general, was henceforth authorized to grant a patent if it considered the invention useful and important. Jefferson, as secretary of state, was the first to administer this law, and sixty-seven patents were granted

[46] 9 Wheaton 1 (1824).

[47] *The Constitution of the United States,* Art. 1, Sec. 8, Par. 8.

[48] House Judiciary Committee, "Patent Act of 1952," *Journal of Patent Office Society,* XXXIV, 553.

under it. The members of the cabinet experienced difficulty in giving careful study to the increasing number of applications. Moreover, their strict standards and their decisions that the states could not grant patents rendered the Act of 1790 unpopular. The year 1793 marks the passage of the second patent law—the only one ever passed in this country which provided for the issuance of patents on inventions without an examination into their novelty and utility. According to this arrangement, known as the registration system, anyone who swore to the originality of his invention and paid the stipulated fees could secure a patent; its validity was decided in the courts. The Act of 1836 repealed the law of 1793 and marked the beginning of our present patent system. It provided for the close scrutiny of each application for patent—in other words, for the examination system. It created the Patent Office, with a commissioner of patents at its head. The law of 1870 was largely a consolidation of the Act of 1836 and supplementary statutes. The Patent Act of 1952 is a second codification of statutes and judicial decisions pertaining to patents. It also contains amendments, some of which are procedural only. Others, particularly the one on contributory infringement, seek important changes. References to this act appear in this and other chapters.[49]

From 1790 to 1836, even with a registration system in effect except for three of those forty-six years, the total number of patents issued was only 9,957. From 1840–49 the number was 5,902, and from 1940–49 it became 351,886. Patent number 1,000,000 was issued in 1911, number 2,000,000 in 1935, only twenty-four years later, and number 2,600,000 in 1952.

Salient Features of Present Patent Law

Patentability. In general three prerequisites—novelty, invention, and utility—must be met in order for one to secure a patent. Novelty, with some exceptions, means that one must be the first to invent. Invention is a contribution over and above the exercise of mechanical skill. The test of utility has not played an important part. The invention may be a process or method, machine, manufacture, composition of matter, or any improvement of them. Also a patent may

[49] U.S. Code Title 35—Patents, Sec. 271.

be conferred for a design which is imposed upon an article of manufacture, or for a new variety of plant. The principal contrast is between a process and a product. And although a single patent invention may be recognized both in the process of making an article and in the article itself, it is a well-settled rule of law that a process patent protects only the process or method and that a product patent protects only the product. The only exception provided by statute is that an unpatented article made in a foreign country by a process that would be infringing if made here cannot be imported.

The subjects for which patents are granted exclude, however, any idea or principle not embodied in concrete form, such as mathematical formulae, laws of nature, and philosophical abstractions. Further, the Atomic Energy Act provided that no patent should be granted after August 1, 1946, on any invention useful solely in the production of "fissionable material" or in the utilization of atomic energy for a military weapon. Additional legislation in 1954 modified the law of 1946 and permits a private enterprise to obtain patents on all atomic inventions except weapons, though competitors may use them on a royalty basis, and requires that findings from government-financed research must become public property.

A person who conceives an invention is entitled to a patent unless he "has abandoned the invention," unless it "was patented or described in a printed publication in this or a foreign country or in public use or on sale in this country, more than one year prior to the date of the application for patent in the United States," or unless other statutory laws are violated.[50] In general, the law seeks to award a patent to the first inventor who uses reasonable diligence in developing or making his invention. There is no time limit between the filing date and the date of issue. The patent itself runs for seventeen years from the date of issue, and there can be no extension beyond that except by act of Congress (after both world wars, in 1928 and 1952, Congress granted an extension to veterans). The names of the inventors appear on all copies of patents and in all patent office records. If an assignment is recorded on or before the payment of the final fee on the patent application, the names of the assignees are printed in the patent copies. Lastly, if the invention is made available to the public by or with the consent of the patentee, he must, by marking

[50] *Ibid.*, Sec. 102.

it or in some other way, give sufficient notice to the public that it is patented.

In the beginning it was easy to fit the definition of invention to the simple economic conditions that prevailed. Any invention was ordinarily the creation of one individual. Since then, however, team research in the laboratory of the large corporation has largely displaced the inventive activity of the individual. The assembly line of invention, like that of manufacturing, is dominant today. The improvements of various workmen and technicians are put together, under the guidance of lawyers and business managers, so that patents can be acquired which will provide for dominating a field of production more completely. Is such a minute division of labor an inventive act? Is the specialized technician along the assembly line of improvement an inventor?

In 1942, Circuit Judge Learned Hand said that invention is not the result of the "inevitable progress of an industry through trial and error" or "of the exercise of persistent and intelligent search for improvement."[51] In 1944, Judge Thurman Arnold stated that a discovery which is the result of step-by-step experimentation does not rise to the level of invention. "Each man is given a section of the hay to search. The man who finds the needle shows no more 'genius' and no more ability than the others who are searching different portions of the haystack."[52] In such organized research "the 'inventor' is paid only a salary, he gets no royalties, he has no property rights in the improvements which he helps to create. To give patents for such routine experimentation on a vast scale is to use the patent law to reward capital investment, and create monopolies for corporate organizers instead of men of inventive genius."[53]

As early as 1850, in *Hotchkiss* v. *Greenwood,* the Supreme Court differentiated the work of the skilled mechanic from that of the inventor.[54] In 1880, in *Densmore* v. *Scofield,* the court employed the expression "a flash of thought" in describing an essential attribute of invention.[55] And in 1941, in the well-known Cuno case, Justice William O. Douglas, speaking for the Supreme Court, declared: "We

[51] *Picard* v. *United Aircraft Corp.,* 128 F. 2d 632, 636 (2nd Cir. 1942).
[52] *Potts* v. *Coe,* 140 F. 2d 470, 475 (D. C. Cir. 1944).
[53] *Ibid.,* 474.
[54] 52 U.S. 248 (1850).
[55] 102 U.S. 358, 375 (1880).

may concede that the functions performed by Mead's combination were new and useful. But that does not necessarily make the device patentable. Under the statute the device must not only be 'new and useful,' it must also be an 'invention' or 'discovery.' Since *Hotchkiss* v. *Greenwood,* decided in 1850, it has been recognized that if an improvement is to obtain the privileged position of a patent, more ingenuity must be involved than the work of a mechanic skilled in the art. . . . The principle of the Hotchkiss case applies to the adaptation or combination of old or well-known devices for new uses. That is to say, the new device, however useful it may be, must reveal the flash of creative genius, not merely the skill of the calling. If it fails, it has not established its right to a private grant on the public domain. . . . Tested by that principle Mead's device was not patentable. We cannot conclude that his skill in making this contribution reached the level of inventive genius which the Constitution authorizes Congress to reward."[56]

In another case in 1945, *Sinclair* v. *Interchemical,* the Supreme Court stressed innovation rather than flash of genius, saying, "Some substantial innovation is necessary, an innovation for which society is truly indebted to the efforts of the patentee. Whether or not those efforts are of a special kind does not concern us. The primary purpose of our patent system is not reward of the individual but the advancement of the arts and sciences. Its inducement is directed to disclosure of advances in knowledge which will be beneficial to society; it is not a certificate of merit, but an incentive to disclosure. Consequently it is not concerned with the quality of the inventor's mind, but with the quality of his product."[57] Justices Hugo L. Black and William O. Douglas, apparently disagreeing with this line of reasoning, "concur in the result."[58] In keeping with this decision, the Patent Act of 1952, Section 103, provides that "Patentability shall not be negatived by the manner in which the invention was made."

Whatever the criteria of patentability may be, it is a significant fact that in 89 per cent of all patent cases brought up to the courts of appeals and the Supreme Court during 1941–45, the patents involved were declared invalid.[59]

[56] *Cuno Engineering Corp.* v. *Automatic Devices Corp.,* 314 U.S. 84, 90–91 (1941).
[57] 325 U.S. 327, 330–31 (1945).
[58] *Ibid.,* 335.

Application for Patent. The application for a patent consists of six parts: the petition, the power of attorney, the specification, the claims, the official drawings, and the oath. The petition is a request to the commissioner that a patent be granted to the applicant. The power of attorney is a legal appointment of the inventor's representative before the Patent Office. The specification describes the invention in connection with the drawings. The claims consist of an enumeration of what the inventor describes as his invention. The official drawings include different views of the invention and its parts from various angles. The oath sets forth the applicant's belief that he is the original and first inventor of the thing for which he solicits a patent. If the invention results from assembly-line research in which many participate, those engaged in such research, including the director of it, must determine which ones can furnish the oath and therefore be considered as joint inventors.

The application goes to one of the sixty-seven divisions of the Patent Office, each of which specializes in a particular group of products. It is taken up for action in the chronological order of filing, an exception being made if assurance is given that an early issue of the patent will result in immediate quantity manufacture of the invention. An examiner scrutinizes it and usually rejects some of the claims. The inventor through his attorney modifies the claims and within six months submits them again. Then the examiner usually makes other objections, and the applicant seeks to remove them. This process may be repeated again and again. In the meantime, according to a time-honored practice which was incorporated into the Patent Act of 1952, the application is kept a secret. If the application is finally rejected, the inventor, upon payment of a fee of twenty-five dollars, may appeal to a board of appeals composed of the commissioner, assistant commissioners, and nine examiners-in-chief, three of whom hear each appeal; then, if necessary, an appeal can be made either to the Court of Customs and Patent Appeals or to the District Court in Washington. About one-half of the patent applications are finally and completely rejected either by the Patent Office or by the courts. Further, one patent application may clash with another or with an unexpired patent because the two, partly or entirely, cover

[59] Daniel G. Cullen, "Patents in Litigation, 1941–45," *Journal of the Patent Office Society,* Dec. 1946, 903–904.

the same invention, whereupon interference proceedings are necessary to determine which one is entitled to a patent. The one who loses the decision may appeal to a court. According to information from the Patent Office, from 1947 to 1952 the percentage of the applications involved in interference proceedings ranged from 0.5 per cent to 0.9 per cent. About 84 per cent of these proceedings were terminated in favor of the first filing. After the issuance of a patent the commissioner can cancel its claims only if the final judgment in an interference proceeding declares that the patentee is not the first inventor of his alleged invention. Fees include one of thirty dollars on filing each original application for a patent, and another of the same amount on issuing each original patent; but subsequently there is no annual fee or tax as in other countries.

The patent law provides for secrecy respecting certain inventions for the sake of national security. The time for taking any action on any government-owned application covering an invention deemed important by a defense agency may be extended to three years. Also, if the invention is one in which the government has a proper interest, the secretary of defense or other government officials may order the commissioner to keep it secret and withhold the grant of a patent. If the commissioner considers the disclosure of any other invention by patent grant detrimental to national security, he must make the application available for inspection to the Atomic Energy Commission, the Department of Defense, or similar agency, any of which may order the commissioner to keep it secret for such period as national security requires. Supplementing this measure are appropriate restrictions upon the application for patents abroad.

Rights of Patentee. The Patent Act of 1952 adopted the language of the Supreme Court in defining a patent as "the right to exclude others from making, using, or selling the invention."[60] In other words, according to this tribunal, it is "a right to be free from competition in the practice of the invention."[61] This right may be transferred by means of an assignment, grant, or license. Any assignment is void as against any subsequent purchaser or mortgagee without notice, unless it is recorded within ninety days. The patentee may appeal to the courts to restrain others from infringing this right—that is, from

[60] Section 154; *Bloomer* v. *McQuewan,* 14 How. 539, 549 (1852).
[61] *Mercoid Corp.* v. *Mid-Continent Investment Co.,* 320 U.S. 661, 665 (1944).

making, using, or selling his invention without his consent. If the government or any of its contractors is the alleged infringer, the patentee may sue in the Court of Claims, or in the case of Tennessee Valley Authority, in the appropriate district court. Until 1953 no one had any occasion to take such action against TVA. According to information received from TVA, in every case in which it has found it desirable to use an invention owned by someone else, it has been able to obtain a license on satisfactory terms. However, according to a recent law, if any defense agency withholds and uses any invention, it may upon request of the patentee, make full settlement for damages. If full agreement is impossible, it may pay 75 per cent of the amount deemed proper, and the applicant may sue for any additional amount.[62] Instead of making such a request for settlement, the patentee may bring suit in the Court of Claims—a highly unsatisfactory remedy because of the delay and expense of litigation. According to the Patent Act of 1952, contributory infringement, in effect, is the sale of a component of an invention or of an essential product especially adapted for use in a patented process.

A charge of infringement of a patent puts in issue either the title of the patentee, the validity of the patent, or the appraisal of the alleged infringement in terms of the claims in the patent. All other rights in the invention must be adjudicated in the same manner as if the patent did not exist.

Relation of Patents to Competition. A patent confers a monopoly on a specific invention. It does not sanction a monopoly of kindred and competitive patents or a restraint of trade other than in the particular thing which the patent covers. The Constitution provides for the granting of monopolies to inventors only "for limited times," and the patent statute is in harmony with this restriction. Patented products may be in competition with both patented and unpatented goods. In fact, the patent law is conducive to competition in that it stimulates individual initiative and private enterprise.

From the beginning there have been patented variations on most products, especially on tools and devices in everyday use, but seldom basic patents on the products as such. Any exclusive control has come generally from bringing all patents in a field under one control; for example, there was competition in the production and sale of parking

[62] United States Code Title 35—Patents, Sec. 183.

meters, all of which were patented, until a promoter brought all of them together. There still remains competition in electrical appliances and many other lines of merchandise. Neither General Electric nor Westinghouse Electric and Manufacturing Company is the leader in the production and sale of washing machines, vacuum cleaners, toasters, waffle irons, or flatirons. Small companies are usually more specialized and aggressive in improving and marketing their products. Furthermore, each of them has its own patents, along with a trademark, which provides a sheltered position from which to compete. Patents may be more helpful to the small manufacturer than the large one, especially in getting started. The beginner may sell his product on the basis of a patented feature rather than price and therefore overcome the handicap of small volume and high cost.

After and perhaps before the expiration of a patent on a basic invention, improvements lead to competition unless the patents on them are concentrated in a pool or single company, or unless one of the owners licenses the others upon restrictive terms. Hundreds of improvement patents have been taken out on refrigeration devices by different companies. Each company owns patented features that others cannot encroach upon without infringement—a situation which has stimulated competition in making as well as in selling a better product. The vice-president of the Packard Motor Company, in reply to a question about the possible effect upon competitors of an important invention that could be used by one manufacturer only, said that "one of the other automobile companies would probably develop something just as good or better in the next few weeks or months, and as a result there would be competition."[63]

The House Committee on Patents reported in 1912 that patents normally tend to stimulate competition. "The patent law was enacted before any of our trusts came into existence; when competition was the accepted law of trade, and when combinations in restraint of trade were not in contemplation. So long as such conditions continued and the avenues of competition were kept open, the monopoly granted by the patent law, limited as it was, in time tended to stimulate competition. It incited inventors to new efforts, and capitalists and businessmen were encouraged to develop inventions. Under these conditions a patent, while granting a monopoly in a specific article, had

[63] TNEC Hearings, Part 2, 321.

rarely a tendency to monopolize any branch of the trade, because few inventions were so fundamental in character as to give the owner of the patent a monopoly in any branch of the trade, and every great financial success arising from an individual patent was sure to result in rival inventions."[64]

Arguments for Patents

According to the Constitution, the general purpose of patents is to promote or stimulate the progress of science and the useful arts. More specific arguments, in keeping with this purpose, should be noted because they provide a point of departure for studying the economic effects of our patent system.

Reward of Inventor. Some have contended that the inventor, as a matter of justice, is entitled to a reward for his contribution to society. In discussing the originator of an improved process and his exclusive privilege of using it for a limited period, John Stuart Mill stated, "That he ought to be both compensated and rewarded for it, will not be denied," and, further, "It would be a gross immorality in the law to set everybody free to use a person's work without his consent, and without giving him an equivalent."[65] Jeremy Bentham asserted, "A patent considered as a recompense for the increase given to the general stock of wealth by an invention, as a recompense for industry and genius and ingenuity, is proportionate and essentially just."[66] A proper appraisal of the necessity of reward, as already indicated in discussing the causes of invention, must differentiate between the scientist, the inventor, and the innovator or improver.

Disclosure of Invention. According to one interpretation, a patent monopoly involves a contractual relationship presumably of mutual benefit to the inventor and the public. The consideration to the inventor is an exclusive monopoly covering his invention for a term of years, while that to the public consists of an immediate and complete disclosure of the invention. Under this arrangement the revealed invention becomes common property after the expiration of the patent. The requirement of disclosure was easily enforced years ago when inventions consisted of contrivances and simple processes. To-

64 Oldfield Report of 1912, 2–3.
65 *Principles of Political Economy,* Bk. 5, Chap. 10, Sec. 4.
66 W. Stark, *Jeremy Bentham's Economic Writings,* 263.

day, however, many patents, especially in the chemical and electrical fields, contain faulty claims. The applicant is no longer required to demonstrate his invention, and the examiner does not try to put it into practice. The examination is devoted not to workability but to similarity of description in previous patents. Vague and incomplete claims would be held invalid by the courts if exposed by the facts. However, only those who can and will spend large sums of money can indulge in the expense and risk of proving insufficient disclosure; and rival companies large enough for mutual respect and understanding seldom challenge each other's patents. Incomplete disclosure may mean in effect that the inventor keeps his invention and yet obtains a patent—he has his cake and eats it too. Stated differently, the public is deprived of the invention and yet grants a patent to the inventor which may be sufficient to keep others away.

The alternative to disclosure, partial or complete, is secrecy. Any process or apparatus the nature or operation of which can be withheld from others, even from employees, invites secrecy rather than disclosure through patents. Manufacturers of certain kinds of chinaware, alcoholic liquids, soft drinks, and foods rely greatly upon secrecy to maintain their pre-eminence. Coca-Cola, for example, has depended upon secrecy rather than a patent for the protection of its process, and its annual profits, for several spans of seventeen years, testify to the wisdom of its choice. In fact, it is doubtful if a patent would have been effective, because it is possible to use and therefore infringe a process without detection. Additional evidence of withholding inventions is found in agreements between companies whereby each will permit the other to acquire "rights to use secret processes."[67] Also, "certain manufacturers feel that it is poor policy to take out patents on production machines" because "the chances are small that some competitor will discover the details of these devices and utilize them."[68]

Certain aspects of our patent system may encourage secrecy in regard to inventions. The requirement that a licensee or laborer assign his patents to others has this effect. Standard Oil Company of New Jersey sought to require users of its hydrogenation process to assign their patents on catalysts in return. A prospective hydrogenation li-

[67] Pooling Hearings, Part 2, 1233.
[68] Comp. L. Hearings, 542.

censee, the Standard Oil Company of Indiana, said, "This would tend to promote the development of secret processes for catalyst manufacture."[69] Moreover, the inventor may fear becoming involved in ruinous litigation in defending his patents.

Development of Invention. Another interpretation of patents, very popular of late in certain quarters, is that they foster the development of inventions. A recent report of the United Nations interprets the grant of patents by various nations from this point of view: "Recognizing that certain conditions of economic security are required in order to encourage investment in what may turn out to be costly (and at times fruitless) research programmes, governments grant to inventors the right to and the benefits of patent monopolies."[70] Development includes the refinement of an idea, the provision of the right kind of machinery and labor, and perhaps the education of prospective customers, and therefore requires capital and risk-taking. Of course competition in the production and sale of the invention is supposed to come after the expiration of the patent. This argument marks a shift from the stimulation of invention to the encouragement of investment—from the reward of the inventor to the reward of the investor.

Patents, in furnishing protection to inventors by excluding others and thus encouraging the development of inventions, are somewhat analogous to a protective tariff which attracts labor and capital to infant industries. Also, patents, like a tariff, may influence the rate of development. They may hasten the perfection and introduction of new inventions because of the limited period in which the patent owners may exploit their legal monopolies. Further, patents, like a high tariff, assist in shifting from an agricultural to an industrial economy.[71] Also, continued insistence upon protection, whether by patents or a tariff, comes mostly from those whose position is threatened.

The established factory, if efficiently operated, has the least need for patents. It has financial reserves, technical know-how, well-known trade-marks, and perhaps a research laboratory. Moreover, it may be subject to high income taxes or public regulation of its rates, and

69 Bone Hearings, Part 7, 3343.

70 *Restrictive Business Practices: Analysis of Governmental Measures Relating to Restrictive Business Practices,* United Nations Economic and Social Council *Report,* 5.

71 Bone Hearings, Part 3, 1272.

therefore can deduct the expense of development in calculating the tax or fair rate of return. Again, it may be a subsidiary and assured of a market, as Western Electric Company is for the telephone equipment it manufactures. Most important, automobile and other manufacturers testify that the greatest incentive in introducing improved and new products is leadership rather than monopoly in the field.

On the other hand, the beginner generally needs patent protection. A manufacturer of automobile parts stated, "The patent affords the parts manufacturer the opportunity to get his initial development and tooling expense back before his competitors start copying his device."[72] George Baekeland said, "The pioneer always has to overcome resistance, has to do the demonstrating, and showing that these things are necessary and useful, and then having done so and established a market, his competitor can come in without any of that pioneering expense. Now a patent of course gives the inventor, the pioneer, the protection which he needs for that pioneering work."[73] However, the so-called patent protection becomes a misnomer if it can be put into effect only through litigation. Much better than that would be a good chance of keeping the invention secret.

Patents may also be helpful in inducing outside capital to develop an invention. According to a manufacturer of automobile parts, "Looking over the large number of parts manufactured by the industry, it is evident that practically all of the companies manufacturing those parts started because of patents. Speculative capital was attracted to these new parts industries when they were new, because investors were convinced that here was something that could be sold in a large volume to the automotive industry, and that they could secure protection long enough to secure the return of their capital and make a fair profit on it. So in the beginning of practically all of these various parts industries the patent was the nucleus around which they were built."[74]

It should be emphasized, however, that every important invention has been developed, patents or no patents; in other words, a patent or an exclusive license under it is not a prerequisite. Recent evidence lends support to this statement. The patented inventions of auto-

[72] TNEC Hearings, Part 3, 1049.
[73] *Ibid.*, 1083.
[74] *Ibid.*, 1049.

mobile manufacturers have been available royalty-free to all, and yet the proportion of them developed and used is apparently as high as for other patented inventions which are not freely available. A manufacturer of automobile wheels said, "When we find there is something that we might infringe, we contact that patentee and we try to get a license. In practically every case we are satisfied with a nonexclusive license, or if he wants to sell the patent for a reasonable amount we might buy the patent."[75]

The German patents impounded during World War I were licensed on equal terms to all.[76] The Office of Alien Property adopted this same policy in administering enemy patents during and after World War II, and the cases in which a policy of nonexclusive licensing might have prevented full exploitation of a patented invention appear extremely rare. It licensed patents on a nonexclusive basis for approximately seven years, and during this period it granted nearly three thousand licenses under 12,677 different patents to over 1,200 different licensees, and, from 1944 to 1946, total production exceeded $200,000,000 in value. Very few license applicants contended that greater production of the patented item would be achieved under an exclusive license, and no applicant made a serious attempt to demonstrate that this would be true in the case of the patents in which it was interested.[77] Two of the largest private organizations conducting research on a nonprofit basis—the Research Corporation and the Chemical Foundation—report success in obtaining industrial exploitation of their inventions under nonexclusive licenses, even where industry originally insisted upon exclusive rights. Likewise, universities which take out patents on research which they sponsor usually issue licenses on a nonexclusive basis. In addition, outstanding inventions within government departments find their way into public use whether exploited exclusively or freely.[78]

Whether or not a company insists that exclusive control of an invention is essential to its development depends upon the line-up of its interests. Standard of New Jersey and a German company, in control of the most important patents on synthetic rubber, advised Goodyear Tire and Rubber Company in 1941 that "they did not be-

[75] *Ibid.*, 1063.

[76] Pooling Hearings, Part I, 740; Bone Hearings, Part 3, 1203.

[77] MP Hearings, Part 2–B, 1459.

[78] *Investigation of Government Patent Practices and Policies,* III, 54.

lieve it was in the interest of the commercial development of this new industry to grant an exclusive license to anyone, but that they hoped to work out some plan under which the entire rubber industry could use the processes on a fair basis."[79] Very likely these two companies would have insisted that exclusive control over their hydrogenated gasoline, the patents on which they held, was essential to its development. Two things appeal strongly to every monopoly: first, its own monopoly; and, second, competition among buyers and sellers with whom it deals.

Public Welfare. In several decisions the Supreme Court has emphasized that the award of a patent is a means to an end. "It is undeniably true, that the limited and temporary monopoly granted to inventors was never designed for their exclusive profit or advantage; the benefit to the public or community at large was another and doubtless the primary object in granting and securing that monopoly."[80] About one hundred years later this same tribunal declared, "It is the public interest which is dominant in the patent system."[81] An opportunity to apply this theory arose when the University of Wisconsin Alumni Association, the assignee of Harry Steenbock's patents on irradiation, refused to license manufacturers of oleomargarine and thus deprived oleomargarine consumers of health-giving vitamins. In an infringement suit involving these patents, a circuit court of appeals declared in 1944 that the inequitable misuse of the monopoly of the patent warrants the denial of equitable relief. "It is now well established that a patentee may not put his property in the patent to a use contra to the public interest."[82] Further, the patent is a privilege "conditioned by a public purpose."[83] This principle has been enunciated also in infringement suits involving restraint of trade.[84]

Does there exist a harmony of interest between a patent system and public welfare? The answer depends upon the evidence which one examines. There are still a few independent inventors who receive a reward from their patents by selling or licensing them. They still have something left after paying the expenses of litigation. Also

79 Bone Hearings, Part 6, 2703.

80 *Kendall* v. *Windsor,* 21 How. 322, 327–28 (1858).

81 *Mercoid Corp.* v. *Mid-Continent Investment Co.,* 320 U.S. 661, 665 (1944).

82 *Vitamin Technologists* v. *Wisconsin Alumni Research Foundation,* 146 F. 2d 941, 944 (9th Cir. 1944).

83 *Ibid.,* 944.

84 See, for example, *Morton Salt Co.* v. *Suppiger Co.,* 314 U.S. 488, 492 (1942).

in certain fields patents are still taken out and used without undue concentration and restraint. Industries that illustrate this situation are those manufacturing washing machines and electric refrigerators and even less important articles such as can openers and other gadgets. Again, there is a patent pool among manufacturers of automobiles—a pool established upon conditions which permit competition. Additional evidence, mostly unfavorable, is presented in later chapters.

Directors of Inventive Activity and Owners of Patents

As previously indicated, those who direct inventive activity and own patents are either independent inventors themselves or business corporations and the government. Their relative importance has changed greatly since World War I.

Inventors. Our patent system was designed to stimulate the individual to invent by giving him the right to exclude others from making, using, and selling his invention. He may exploit this right himself or sell it to someone else. He invents largely according to his inclination, often roaming into bypaths with little regard to commercial values—an environment conducive to worthless contraptions as well as ingenious ideas. From him has come most of the great inventions. Since the turn of the century he has sold his patents mostly to business corporations. If he becomes an employee of a business corporation or the government, he is usually required to assign his inventions to them without additional compensation.

Business Corporations. About the turn of the century a few large corporations, such as General Electric and American Telephone and Telegraph Company, were created by consolidating the physical assets and patents of other companies. The work of their engineers and technicians in developing their products and services led to the research laboratories which they now have. Today dozens of other corporations have similar facilities.

The research of any corporation lies in its particular field and its primary purpose is to solve immediate technical problems. Another objective is to anticipate inventive trends and take out patents to keep open the road of technical progress and business expansion. Coca-Cola, for example, maintains a research staff in the soft drink field. It neither collects royalties from its patents on bottling and other facilities nor restricts their use.[85] Also patents give bargaining power in cross-licensing and other agreements. Further, they may pre-empt alternate products and methods and bring about the monopoly of an industry, and also permit the timing of improvements to minimize obsolescence. Another inducement is that research is a deductible expense in the reckoning of income taxes.

The research of corporations may lead into bypaths and by-products, a fact illustrated by American Telephone and Telegraph's development of X-ray apparatus and hearing aids. However, great inventions, such as the radio and fluorescent lighting, generally come from the outside. A few of them, such as cellophane, have been developed in industrial laboratories.

In some industries the members pool their patents and each acquires the right to use all of them. The pool may also acquire patents from outsiders. Most important is the practice of a corporation in acquiring patents from others through purchase and exclusive license and thus building up and maintaining a dominant position. The Supreme Court has not yet ruled upon the legality of consolidating all patents in one industry.

The hired inventors and technicians of business corporations generally receive only salaries for compensation, the patents on their inventions being assigned to their employers. According to an investigation of three thousand consecutively-numbered patents of 1937, 53 per cent were assigned to corporations.[86] They came from employees as well as outside inventors. Since World War I, business corporations have conducted much research financed by the government, and they generally have kept the patents on the thousands of inventions resulting from such work.

Government. For several decades various departments of the government have carried on various lines of research and fostered invention. Because of both administrative policy and judicial interpretation, the government has received until lately only shop rights in most of the inventions of its employees—the right to use them without royalty in providing its military and other equipment. With only this qualification the Army and Navy and a few other departments let their inventors take out patents and sell their patent rights to others.

85 MP Hearings, Part I, 421.
86 Comp. L. Hearings, 306.

Further, the Supreme Court, in 1933, declared that in the absence of contract the government, like any employer under common law, is legally entitled only to the inventions which its employees are hired to make.[87] These two restrictions, one of policy and the other of interpretation, have resulted in the government becoming the owner of comparatively few patents. In 1950, a presidential order directed the government to obtain the entire right to inventions made by its employees during working hours; made with its substantial contributions in the form of facilities, equipment, materials, funds, information, or help of other employees; or made in consequence of official duty.

The Atomic Energy Commission, created in 1946, is the outstanding example of a specific field of research under the immediate direction of the government. As previously mentioned, no patent may be granted to others on any invention useful solely in the production of fissionable material or in the utilization of atomic energy for a military weapon. This commission, by August, 1954, had released a total of 704 patents for licensing. Nearly 400 licenses to use these patents have been issued. According to information from the AEC, a number of invention disclosures have been made to the Commission but whether they are patentable or not is a moot question until they actually are patented. As already mentioned, a law of 1954 relaxes somewhat the patent provisions of the Atomic Energy Act of 1946.

In 1947, Congress created the Research and Development Board to assist the secretary of defense with a program of research and development for military purposes—a program which includes the government, universities, and business corporations. It directs its efforts toward the best application of science and technology to achieve the most effective weapons and countermeasures. It is not an operating agency and therefore does not let contracts or directly administer projects. It consists of a civilian chairman and two members each from the army, navy, and air force. A Munitions Board advises concerning military production plans and materials in short supply.[88]

Several official studies—for example "Research—A National Resource" in 1938, "Science, The Endless Frontier" in 1945, and "Science and Public Policy" in 1947—proposed that the government

[87] *United States* v. *Dubilier Condenser Corp.*, 289 U.S. 178 (1933).

[88] *Research and Development Board: Structure and Functions*, U. S. Department of Defense Report (1952).

achieve a balanced and more effective science program for the nation. In response to this, Congress in 1950 created the National Science Foundation. Its general purpose is to perform basic research in the mathematical, physical, medical, biological, engineering, and other sciences, and, upon request, to perform specific research relating to national defense. One of its first acts was the award of four hundred fellowships to the most meritorious of a much larger number of applicants. Each contract awarded the Foundation for scientific research must provide for the disposition of resultant inventions "in a manner calculated to protect the public interest and the equities of the individual or organization" with which it is executed—a requirement flexible enough to permit either the Foundation or the research corporation to become the owner of the patents. This requirement, in contrast with the patent provisions of the original bills, is a concession to industrial enterprises. Private business opposed the earlier bills because they directed that patent rights be left with the government. Other objectives of the Foundation include the interchange of scientific information, correlation of research programs, and a clearing house for research personnel.

There is considerable co-operation among the AEC, RDB, and NSF in the field of research. The objective is to assure a balanced research program, without either gaps or duplication of effort. The NSF's fellowship program, for example, is replacing the AEC's in the general physical and biological fields.

In 1930, the total expenditures for research by private corporations and the government were about $166,000,000, and by 1940 they were $345,000,000. In 1952, the total was three billions, nearly two-thirds of which was provided by the federal government.[89] However, the proportion of research cost borne by the federal government varies widely by industry. In 1951, for example, the government provided the funds for 85 per cent of the research performed in the aircraft industry and for about 50 per cent of that in electrical machinery but only 7 per cent in the chemical industry and 3 per cent in petroleum refining.[90]

Professors of universities which are supported partly by the appro-

89 National Science Foundation Annual *Report* (1952), 6.

90 *Industrial Research and Development*, U. S. Department of Labor and U. S. Department of Defense *Joint Report* (1953), 10.

priations of states and cities have been the source of important inventions. Professors Henry of Princeton and Babcock of Wisconsin have been mentioned already. Professor F. G. Banting developed the use of insulin for diabetes and gave the patents to the University of Toronto, which then licensed drug manufacturers to make and sell it. There are other notable examples in Europe as well as in America. The research of professors, in proportion to its cost, is the most fruitful, not only in findings but in reaching publication.

A minor field of government activity is the helping to find a market for inventions. The National Inventors Council, created in 1940 by the secretary of commerce, functions in an advisory capacity to the government in appraising inventions for the national defense and welfare. In addition it assists the public in submitting such inventions, acquaints them with problems confronting the armed services, and refers particularly valuable ideas to the Department of Defense or other appropriate agencies. It limits its activity to placing the inventor in contact with proper authorities if his proposal appears to meet a government need. Appropriate agencies, such as the army and navy and air force, may "make arrangements with the inventor for the use of his invention."[91] This is contingent of course upon their approval and adoption of the invention for immediate or future use. The membership of the National Inventor's Council, eighteen in number, includes outstanding inventors, scientists, and industrial research experts with experience in developing and using inventions. During World War II this organization screened many inventions for the armed services, and in 1951 it again became active.

Conclusion

Invention is a response to the general environment—an environment which provides an opportunity for the instinct of contrivance to express itself, offers the esteem of others, manages the division of labor, honors intellectual speculation, furnishes accumulated knowledge and suitable economic conditions, and proffers economic reward. Patents furnish one possible means by which the inventor can obtain his reward and, to that extent therefore, patents promote the industrial

[91] *National Inventors Council,* U. S. Department of Commerce *Information Bulletin* (1951), 3.

arts and economic welfare of the nation. A salient feature of the present administration of the patent law, however, is that the judicial concept of patentability differs from that of the Patent Office to such an extent that a high proportion of the adjudicated patents are held invalid. Furthermore, one of the most acceptable arguments on behalf of patents—the disclosure of invention—has been weakened by the failure in important instances to reveal the necessary technology for reducing the invention to practice.

CHAPTER II

Patent Pools

TOWARD THE CLOSE of the nineteenth century the United States experienced an increase in the size of factories and other plants and an even greater increase in the concentration of their control. According to the record, this concentration was accomplished and still continues by means of agreements, trusts, pools, holding companies, consolidations, and community of interest through intercorporate stockholding, interlocking directorates, and common stockholders. As the present and the next three chapters point out, this concentration of control was—and still is—in patents as well as in physical assets.

Perhaps the most important action by Congress in curbing any restraint of trade and monopoly by these means was the passage of the Sherman Act in 1890. Another law, which had already been in operation for many years, was the patent statute conferring upon the inventor the exclusive right to make, use, and sell his invention for a period of years. Where does one of these laws begin to operate and where does the other end? Suppose that kindred patents are brought together by agreement or purchase and that their rights are exercised collectively rather than individually—is the aggregate, legally speaking, the mere sum of the parts? If the patentee controls the output and use and price of his invention through his licensees, and if these licensees are former competitors, is the effect a violation of the Sherman Act or a mere exercise of patent rights? The settlement of these and kindred questions by the courts provides the subject matter for most of this book.

Before proceeding with cases, the general characteristics of pools should be noted. A pool as a form of business organization is a means by which various companies or individuals "pool" or bring together their patents, markets, traffic, revenues, or some other phase of their operations, and divide one or more of them among themselves on some acceptable basis. A patent pool is an arrangement by which two or more patent owners put their patents together and receive in return a license to use them. By means of assignment the control of the patents is usually vested in an individual or a corporation—an arrangement somewhat analogous to the old "trust" in which stockholders of various companies would assign their stock to a board of trustees.

Cross-licensing, usually more direct and limited, may accomplish the same results as a pool. Moreover, a few large companies may, in effect, pool or cross-license their patents simply by taking no action against each other for possible infringement.

In some instances the members pay royalties to the holder of the patents which are later divided among them according to their appraisal of the relative importance of their patents. If they cross-license each other, they make such payments directly to each other. In most cases, however, they pay nothing—obviously a much simpler arrangement. The members may exchange technological information or know-how as well as patent rights. Lastly, the participants in the pool may be limited to the contributors of patents; in some instances, however, others are licensed, and any royalty payments are divided among the patent owners.

One reason for patent pools, one that has generally been approved by the courts, is to provide for the use of complementary and perhaps conflicting inventions. Another reason or pretext may be to avoid litigation about infringement and validity, actual or threatened. Also, according to government reports and court decisions, another purpose may be to restrain trade. The judicial history of patent pools from this standpoint indicates three periods: the period of disregard or even approval of any restraint of trade resulting from patent pools (1789–1909); the period of disapproval of such restraints (1909–42); and the period (after 1942) of continued disapproval, strengthened by the requirement of compulsory licenses with royalties, as the result of court decisions, and by the same requirement with or without royalties, as the result of consent decrees. Each period will be discussed and illustrated in terms of actual cases.

Approval of Pools That Restrain Trade, 1789 to 1909

The first patent pool among manufacturers apparently was that of sewing machine patents in 1856. No legal action, it seems, was taken against it. About the turn of the century patent pools were created in the harrow industry and seeded raisin industry, too, for the primary purpose of restraining trade, and yet federal courts did not condemn them. At this time the courts seemed to think of a patent and its privileges primarily in terms of the individual inventor and his specific product or process—a point of view generally justified by existing conditions. Even the corporations of that day were predominantly small and often promoted by an inventor or his associate who sought to make and sell his patented invention. The judges thought mainly of the inventor and his rights and were prone to ignore any resultant restraint of trade that unduly extended the patent monopoly.

Sewing Machine. The sewing machine industry started in the United States about 1850. It was not very profitable at first owing to the expense of perfecting the machine and introducing it to the public. In 1856, the manufacturers of sewing machines agreed to pool their patents for the purpose of more effectually protecting the business against infringers and of granting licenses under the patents to others who desired to engage in the business. The license fee was fixed at fifteen dollars per machine. In the year 1860, this license was reduced to seven, and at this rate twelve additional licenses to manufacture were granted. In 1868, it was again reduced to five and lastly, in 1870, to three.[1] The members of the pool were Elias Howe, Wheeler and Wilson, Grover and Baker, and I. M. Singer. They dominated the field until after 1877, when the majority of the basic patents expired.

Harrows. The National Harrow Company of New York was organized by six manufacturers of harrows for the purpose of holding patents and licensing the respective manufacturers under them. It acquired control of most of the spring-tooth harrow business in the United States. Bement and Sons was one of the manufacturers of harrows that assigned its patents to this company. The former obtained, under definite conditions, an exclusive license to make the harrows described in the patents which it assigned. Bement was required to pay a royalty, to furnish monthly reports, to make and sell only that type of harrow covered by its patents, to sell harrows only at stipulated prices, and to pay five dollars for each harrow sold contrary to these terms. National Harrow reserved the right to decrease

[1] John S. Perry, *A Defense of the Patent System of the United States*, 24–25.

the selling price or the royalty and to furnish counsel in any infringement suit. It accused Bement of violating these conditions and demanded a large amount of damages and a specific performance of the contract. The defendant denied many of the charges and maintained that since it and the plaintiff, together with other manufacturers of harrows, were members of a combination to regulate the output and price of harrows, their contract was null and void. The New York Supreme Court, an intermediate court, held the license contracts void as against public policy and in contravention of the New York antitrust law.[2]

The case was appealed to the United States Supreme Court in 1902. The evidence, examined by a referee, failed to show either that other manufacturers of harrows entered into the same kind of contract with National Harrow or that a general combination among the dealers in patented harrows existed to regulate the prices of this commodity. The court declined to assume that "there was a general combination among the dealers in patented harrows to regulate the sale and prices of such harrows." It stated that these contracts "are to be judged by their own contents alone and construed accordingly."[3] Moreover, "the plaintiff, according to the finding of the referee, was at the time when these licenses were executed the absolute owner of the letters patents relating to the float spring-tooth harrow business."[4] The general rule is that any conditions "imposed by the patentee and agreed to by the licensee for the right to manufacture or use or sell the article, will be upheld by the courts."[5]

The whole agreement, therefore, between National Harrow and Bement, as interpreted by the court, was merely a contract between a patentee or licensor and a licensee; the patentee, enjoying the legal right to monopolize his patented commodity, may license another to manufacture, use, and sell this commodity upon restrictive conditions. This decision, handed down in 1902, has been improperly cited in many instances in support of price and other restrictive agreements between a patent owner and a group of licensees.

Earlier judicial decisions involved the National Harrow Company of New Jersey. It was a holding company and preceded National

2 *National Harrow Co.* v. *E. Bement and Sons,* 47 N.Y.S. 462 (1897).
3 *Bement* v. *National Harrow Co.,* 186 U.S. 70, 85 (1902).
4 *Ibid.,* 88.
5 *Ibid.,* 91.

Harrow of New York. The operation and purpose of the two companies, however, were apparently identical. In one of these earlier judicial decisions, in 1897, a lower court declared that the purpose "was to form a combination between the various manufacturers of these harrows, to prevent competition in business and enhance prices; and such is the effect of their agreement. The corporation, provided to hold the legal title of the several patents, is merely an instrument to effect this object. . . . The fact that the property involved is covered by letters patent is urged as a justification: but we do not see how any importance can be attributed to this fact."[6]

Seeded Raisins. In 1900, eight manufacturers engaged in seeding and processing raisins assigned their respective patents relating to this art to a corporate dummy and received in return uniform licenses based upon all of the patents. The licenses declared this dummy the owner of the patents and required it to inspect the books of the licensees, to prosecute infringers, to collect royalties, to prohibit the use of other machines by the licensees; and to refrain from licensing others without the consent of one-half the licensees. One of the manufacturers, charged with a violation of the license agreement, pleaded the illegality of the contract. A circuit court of appeals held that such a contract is not void as against public policy.[7] It based its argument upon the Bement decision; but in that case, only the relation of a patentee to a single licensee entered into the argument. Therefore it was illogical to employ that decision as a criterion in the present case —a case in which several owners of patents for similar inventions assigned them to a single company and received in return restrictive licenses based upon all patents thus pooled.

Disapproval of Pools That Restrain Trade, 1909–42

As business units became larger, patent-owning corporations supplanted inventors in the exploitation of patents. Increase in the size of businesses usually resulted in a reduction in the number of enterprises in each industry and thus facilitated the pooling of patents with price and other restrictions. The courts realized, as they had not realized before, that this arrangement was primarily a violation

[6] *National Harrow Co.* v. *Hench,* 83 Fed. 36, 38 (3rd Cir. 1897).

[7] *United States Consolidated Seeded Raisin Co.* v. *Griffin and Skelley Co.,* 126 Fed. 364 (9th Cir. 1903).

of the Sherman Act rather than a mere exercise of patent rights, and in response to vigorous prosecution, they condemned not only restrictive patent pools but also other kinds of monopolies and restraints of trade. This new attitude of the courts is seen, for instance, in the American Tobacco Company and the Standard Oil decisions of 1911.

Liquid Door Check. In the so-called lock case of 1909 one finds an excellent exposition of the patent and Sherman laws and of their relation to each other and to the public. Four companies that owned patents on liquid door checks entered into contracts which restrained each other in the exercise of rights conferred by their respective patents and authorized each of them to use the patents of the others. The plan comprehended, in the language of a circuit court, "the maintaining of prices, the pooling of profits, the elimination of competition, and the restraint of improvements. . . . The principal question is whether the fact that the articles to which the agreements relate embody patented inventions is sufficient to make the Sherman Act inapplicable."[8]

The court declared that inasmuch as "the monopoly granted by letters patent is of a particular invention" which is, "as a matter of fact, in commercial competition with both patented and unpatented devices," a restraint of trade by a combination of independent manufacturers, even though letters patent protected their respective commodities, was not warranted.[9] In other words, a combination cannot be lawful merely because its constituent parts are legal monopolies. The court also stressed the supplementary character of the patent and the Sherman laws. Each has a limited field. The one grants a certain kind of monopoly but not everything that the other forbids. In this instance each of the patentees passed beyond its right to prevent others from making, using, and vending its invention—all that the patent statute confers—by combining or pooling patents in restraint of trade and hence violating the Sherman Act.

Bathtubs. The Standard Sanitary Manufacturing Company, which in 1912 controlled about one-half the output of enameled ware in the United States, and two other manufacturers of enameled iron ware—J. L. Mott Iron Works and L. Wolff Manufacturing Company—assigned their patents for an enameling process to the secretary of

[8] *Blount Mfg. Co.* v. *Yale and Towne Mfg. Co.,* 166 Fed. 555, 556 (C. C. Mass., 1909).
[9] *Ibid.,* 557.

the Association of Enameled Ware Manufacturers. On the basis of the patent assigned by Standard Sanitary, the other patents being regarded as infringing, the secretary issued licenses to manufacturers of enameled ware and their jobbers, bringing 85 per cent of the manufacturers and 90 per cent of the jobbers into a network of license agreements. According to these contracts, a committee of six, chosen from the manufacturers entering the combination, determined the selling prices of both manufacturers and jobbers. The three manufacturers could sell their enameled ware only to those jobbers belonging to the combination; while the jobbers could buy such goods only from the three manufacturers. In addition, these two groups—manufacturers and jobbers—were parties to an agreement approved by the secretary of the association. He was the central figure in a network of licenses that cemented enameled ware manufacturers and jobbers into a strong horizontal and vertical combination. A lower court disapproved the combination for these reasons: each defendant entered into it with knowledge that others had done so and higher prices resulted; the patent in question covered only a minor part of the manufacturing process; and the finished product itself was unpatented.[10]

The Supreme Court, to which the case was appealed, concluded that this monopoly arose primarily from combination and hence condemned it as it had other combinations that had violated the Sherman Act. "The added element of the patent in the case at bar cannot confer immunity from a like condemnation, for the reasons we have stated. And this we say without entering into the consideration of the distinction of rights for which the Government contends between a patented article and a patented tool used in the manufacture of an unpatented article. Rights conferred by patents are indeed very definite and extensive, but they do not give any more than other rights an universal license against positive prohibitions. The Sherman law is a limitation of rights, rights which may be pushed to evil consequences and therefore restrained."[11] It cited the American Tobacco Company decision and other antitrust opinions not involving patents and therefore placed this combination in the category to which other unlawful monopolies belong.

[10] *United States* v. *Standard Sanitary Mfg. Co.,* 191 Fed. 172 (C. C. Md. 1911).
[11] *Standard Sanitary Mfg. Co.* v. *United States,* 226 U.S. 20, 49 (1912).

Motion Pictures. The motion-picture industry arose from pioneer inventions near the end of the nineteenth century. Ten manufacturers of motion-picture films sought to dominate the motion-picture business, including its three phases of manufacture, distribution, and exhibition. The subterfuge consisted of a patent pool. The ten manufacturers organized the Motion Picture Patents Company, to which four of them assigned several patents. These patents were used by this specially organized company as a basis for interlocking licenses, granted not only to manufacturers of motion-picture films and projecting machines, but also to distributors and exhibitors of these films. By leasing, instead of selling films, and by enforcing stringent restrictions and conditions pertaining to them, the control of the motion-picture industry seemed assured. Competition was practically eliminated both from within and without the combination; for example, no two distributors could supply films to the same exhibitor, and the various strata of the industry were compelled to use all or none of the machines and films of the monopoly. Moreover, contracts, royalties, penalties, and the like—all prescribed by this company—were uniform as between any two of the participants in the monopolistic hierarchy of company, manufacturers, distributors, and exhibitors. A manufacturer could not distribute or exhibit films, a distributor could not make or exhibit films, and an exhibitor could not make or distribute films—functional specialization was brought about within the industry.

The government brought suit against this monopoly for violation of the Sherman Act. A district court ruled, "The end and purpose of the combination, and in this sense the motive or moving cause, further was not to protect the patent rights, which the Motion Picture Patents Company was organized to take over; but the control of the patents was made a feature of the scheme, in the belief, or at least the hope, that this would render the scheme (otherwise illegal) not open to the condemnation of the law."[12]

Coaster Brake. The New Departure Manufacturing Company and five other corporations were separately engaged in manufacturing bicycle and motorcycle coaster brakes and accessories of a type and design distinctive from those made by other concerns. The six corporations were manufacturing under certain patent license rights owned

[12] *United States* v. *Motion Picture Patents Co.,* 225 Fed. 800, 811 (E. D. Pa. 1915).

by them separately. By granting licenses to competitors, they sought to restrain trade in coaster brakes. A district court declared that "by the methods and acts complained of, the commingling of separate interests in separate patent rights, by the issuing of a pretended license for a pretended basic patent, with the intention of fixing uniform prices and discounts and imposing other conditions, benefits and advantages were secured which may be enjoyed only by a separate patentee in the protection of his true monopoly. It was such courses of procedure by industrial interests in whatever form or guise that the Antitrust Act was designed to check and prevent."[13]

Automobile Bumpers. Various manufacturers of automobile bumpers organized the Automobile Bumper Association and assigned to it their relevant patents. The association in turn, according to a government complaint, issued to members identical licenses to manufacture bumpers, with royalties a fixed percentage of gross sales, and with provisions as to prices. A consent decree in 1917 dissolved the association and enjoined its further operation. The court stipulated, however, that any member who owned a patent could, "so long as he acts separately and independently," issue any lawful license to "one or more of the defendants acting separately and independently of each other."[14] A consent decree generally represents the minimum which the Department of Justice is willing to accept in lieu of a court decision. A defendant that accepts it considers its terms better than the expense of litigation and perhaps the unfavorable decision that would result from its rejection. A consent decree is granted by a court in place of a decision and is based upon the agreement already reached between the government and the defendant in question.

Gasoline and Kindred Derivatives. At first, gasoline was manufactured from crude oil by distillation and condensation at atmospheric pressure. Efforts to increase the yield resulted in a process for "cracking" the residue—subjecting it to great pressure and heat, thereby releasing the lighter elements and thus transforming them into gasoline. This cracking process was first perfected in 1913 by Standard Oil of Indiana. During the next seven years this company manufactured cracked gasoline on a large scale and also licensed fifteen independent concerns to use its process in return for $15,000,000 in royal-

[13] *United States* v. *New Departure Mfg. Co.,* 204 Fed. 107, 115 (W.D.N.Y. 1913).
[14] *United States* v. *Descher,* I D & J 645, 646–47 (1917).

ties. In the meantime the Texas Company, Standard Oil of New Jersey, and Gasoline Products Company sought independently their own commercial processes for cracking. Beginning in 1920, conflict developed among the four companies concerning the validity, scope, and ownership of their patents and inventions. As a result of the conflict, they pooled their patents professedly to avoid conflict, litigation, and loss. Their agreements released each of them from liability for any past infringement of patents of the others. Each of them could license independent concerns under its process and also release independent concerns from infringement of patents controlled by the others. Each shared in some fixed proportion the fees received under these multiple licenses.[15]

In 1924, the government brought suit against these four companies for violation of the Sherman Act. Seven years later the Supreme Court said: "An interchange of patent rights and a division of royalties according to the value attributed by the parties to their respective patent claims are frequently necessary if technical advancement is not to be blocked by threatened litigation. If the available advantages are open on reasonable terms to all manufacturers desiring to participate, such interchange may promote rather than restrain competition."[16] In reply to the contention that onerous royalties made the arrangement unlawful, the tribunal went on to say, "Unless the industry is dominated, or interstate commerce directly restrained, the Sherman Act does not require cross-licensing patentees to license at reasonable rates others engaged in interstate commerce."[17] Furthermore, "If combining patent owners effectively dominate an industry, the power to fix and maintain royalties is tantamount to the power to fix prices. Where domination exists, a pooling of competing process patents, or an exchange of licenses for the purpose of curtailing the manufacture and supply of an unpatented product, is beyond the privileges conferred by the patents and constitutes a violation of the Sherman Act."[18] Note the emphasis on the significance of dominance. Finally, "But an agreement for cross-licensing and division of royalties violates the Sherman Act only when used to effect a monopoly,

[15] *Standard Oil Co. (Ind.)* v. *United States,* 283 U.S. 163, 167–68 (1931).

[16] *Ibid.,* 171.

[17] *Ibid.,* 172.

[18] *Ibid.,* 174.

or to fix prices, or to impose otherwise an unreasonable restraint upon interstate commerce."[19] The court held that no monopoly resulted in this instance because the pool controlled only 55 per cent of the output of cracked gasoline which in turn was only 26 per cent of the total gasoline production.[20] But is not 26 per cent enough to establish price leadership and even dominance in some localities if the entire business is not evenly distributed?

Other parts of the agreements between these four companies were not free from monopoly or restraint of trade, but the Supreme Court practically ignored them because they had been eliminated. For example, according to one agreement and the licenses based on it, Standard of Indiana "secured an option to purchase a portion of the cracked gasoline manufactured in, or shipped into, its sales territory."[21] Upon the request of a district court the objectionable clauses were voluntarily cancelled. Similarly, the propriety of certain blanket acknowledgements of patent validity by these four companies and their licensees was questioned by the lower court. At the court's suggestion, these provisions also were formally cancelled.[22] The Supreme Court did not consider, therefore, any of the issues concerning the validity or scope of the cracking patents.

During the thirties there were several other patent pools relating to gasoline and kindred derivatives. Standard of New Jersey and Standard of Indiana, along with Atlantic Refining Company and Petroleum Distillation Corporation, pooled their fractional distillation patent rights and in turn received licenses under such patent rights which they owned or might control prior to 1940.[23] In another instance, the two Standard companies, Phillips Petroleum Company, and Texas Company, placed their polymerization patents together and received in return nontransferable licenses based upon them and kindred patents on inventions made prior to July 1, 1947.[24] Again, the two Standard companies and Union Oil Company independently developed certain refining processes for removing asphalt and wax; interference proceedings and the likelihood of infringement led them

19 *Ibid.*, 175.
20 *Ibid.*, 175–76.
21 *Ibid.*, 181.
22 *Ibid.*, 181–82.
23 Pooling Hearings, Part 4, 3698–99.
24 *Ibid.*, 3701–3703.

to cross-license each other under their respective patents on these processes. They agreed to share in the cost of acquiring and maintaining patent rights and to co-operate in a joint program of research and development. Also they authorized another company to receive royalties paid by licensees under their patent rights in this field.[25] Lastly, four companies—Texas, Standard of Indiana, Standard Oil Development, and Pure Oil—acquired in 1933 the patents of the Gray Processes Corporation covering clay treating of naphthas and kerosenes, delivered to the Gray Corporation their own patents in this field, and authorized it to grant each of them a nontransferable license based upon these and kindred patents on inventions made prior to January 1, 1950.[26]

Radio Loudspeakers. In interpreting the cross-licensing of patents by manufacturers of radio loudspeakers, a circuit court of appeals declared, in the light of the oil-cracking case, "If the purpose of a plan is to effect a monopoly, or to fix prices, or to impose otherwise an unreasonable restraint, then the [Sherman] Act is violated by use of any means to that end, including a cross-licensing agreement."[27]

Continued Disapproval of Restraints of Trade and Requirement of Compulsory Licenses after 1942

In the forties, the Department of Justice and Supreme Court began to enforce the Sherman Act in a more effective manner, particularly in those violations involving patents. Their decisions not only condemned monopoly and restraint, but permitted companies not holding particular patents to become licensees at reasonable royalties. In some cases, perhaps the most flagrant, consent decrees have authorized royalty-free licenses and even cancelled patents.

Electrical Fuse Cutout—Line Material Case. Southern States Equipment Corporation owned a dominant patent on an electric fuse cutout, used in preventing short circuit or overload, and the Line Material Company owned an improvement patent on the same product. They were in interference with each other in the Patent Office even prior to the issue of their respective patents, and only when both patents were lawfully used by a single maker could the public

25 *Ibid.*, 3704–17.

26 *Ibid.*, 3733–42.

27 *Lynch* v. *Magnavox Co.*, 94 F. 2d 883, 890 (9th Cir. 1938).

or the patentees obtain full benefit of the efficiency and economy of the inventions. In 1940, each company agreed to grant the other a royalty-free cross-license.

The government brought action against these two companies and their licensees. It did not question the validity of the patents. According to the Supreme Court in 1948, the ultimate question was whether "two or more patentees in the same patent field may legally combine their valid patent monopolies to secure mutual benefits for themselves through contractual agreements, between themselves and other licensees, for control of the sale price of the patented devices."[28] Its reply was that "no case of this Court has construed the patent and antimonopoly statutes to permit separate owners of separate patents by cross-licenses or other arrangements to fix the prices to be charged by them and their licensees for their respective products. Where two or more patentees with competitive non-infringing patents combine them and fix prices on all devices produced under any of the patents, competition is impeded to a greater degree than where a single patentee fixes prices for his licensees. The struggle for profit is less acute. Even when, as here, the devices are not commercially competitive because the subservient patent cannot be practiced without consent of the dominant, the statement holds good. . . . The merging of the benefits of price-fixing under the patents restrains trade in violation of the Sherman Act in the same way as would the fixing of prices between producers of non-patentable goods."[29] The Supreme Court felt in this decision that "when patentees join in an agreement as here to maintain prices on their several products, that agreement, however advantageous it may be to stimulate the broader use of patents, is unlawful *per se* under the Sherman Act. It is more than an exploitation of patents. There is the vice that patentees have combined to fix prices on patented products. It is not the cross-licensing to promote efficient production which is unlawful. There is nothing unlawful in the requirement that a licensee should pay a royalty to compensate the patentee for the invention and the use of the patent. The unlawful element is the use of the control that such cross-licensing gives to fix prices."[30] The lower court, in keeping with this opinion, permitted

28 *United States* v. *Line Material Co.*, 333 U.S. 287, 305 (1948).
29 *Ibid.*, 311.
30 *Ibid.*, 314–15.

the licensees to cancel their contracts, allowed them to obtain licenses under pertinent patents with reasonable royalties, and enjoined any suit for alleged infringement prior to 1948. Another significant aspect of this case—maintenance of resale price—is presented in Chapter Four.

Variable Condensers. Variable condensers are tuning devices used on radios to select broadcasting stations. Beginning in 1934, the only patents pertaining to them were on parts and methods of assembling. In that year three manufacturers, two of which accounted for nearly three-fourths of the output, organized General Instrument Corporation to which they assigned their forty-seven patents and from which they received licenses. They also authorized it to obtain the present and future patents from others and to sue for alleged infringement of its patents. It could grant licenses to others if the members of the pool gave their unanimous approval. Infringement suits against other companies terminated in the grant of price-fixing licenses to them. In 1939, two members paid the third member $50,000 for its patents relating to tuning devices and for its agreement to stay out of business for ten years and to sell its tools and dies and gigs outside the United States and Canada—"naked agreements," according to a district court in 1949, "between competitors designed to restrict the tools and equipment from falling into the open market."[31] This court held that the members of this pool are not in the position of the defendants in the oil-cracking case of 1931. "Rather are they within the contemplation of the court in the same case where it said: 'The cost of royalties may, of course, be a decisive factor in the cost of production. If combining patent owners effectively dominate an industry, the power to fix and maintain royalties is tantamount to the power to fix prices.' "[32] A final judgment in 1953 forbade, among other things, the joint acquisition of patents and agreements on validity of future patents and provided for the dissolution of General Instrument and for compulsory licensing of their patents at reasonable royalties until 1958.[33]

Other Products. Additional patent pools are mentioned in the next three chapters. General Electric and Westinghouse, for example,

[31] *United States* v. *General Instrument Corp.*, 87 F. Supp. 157, 194 (D.N.J. 1949).
[32] *Ibid.*, 195; *Standard Oil Co. (Ind.)* v. *United States*, 283 U.S. 163, 174 (1931).
[33] *United States* v. *General Instrument Corp.*, 115 F. Supp. 582 (D.N.J. 1953).

cross-licensed each other under their patents on incandescent lamps.[34] Also some consolidations, such as the ones in radio and synthetic rubber, though brought about by the outright acquisition of patents, are sometimes incorrectly referred to as pools. Other instances of pools may be found in government complaints and consent decrees.

One law dictionary describes a consent decree or judgment as follows: "One entered by consent of the parties; it is not properly a judicial sentence, but is in the nature of a solemn contract or agreement of the parties, made under the sanction of the court, and in effect an admission by them that the decree is a just determination of their rights upon the real facts of the case, if such facts had been proved. It binds only the consenting parties; and is not binding upon the court."[35]

On motion by the antitrust division of the Department of Justice to vacate a decree involving Radio Corporation of America, a circuit judge ruled in 1942, "A consent decree, although based upon an agreement of the parties rather than a finding of facts by the court, is not a mere authentication or recording of that agreement. It is a judicial act, and, therefore, involves a determination by the chancellor that it is equitable and in the public interest."[36]

The assistant attorney general in charge of the antitrust division of the Department of Justice testified in 1950 about the way a consent decree or judgment comes about: "The government, after investigation, files in court a civil complaint charging the defendants with violation of the antitrust laws. The defendants, if they wish, may follow the formal litigating process, answer the complaint, and try the case. If, on the other hand, they wish to settle the case, they then approach the government and offer a proposed judgment. There follows extensive discussion and negotiation between the government and the defendants. Sometimes this results in a consent judgment. Sometimes we believe that the relief offered by the defendants is not adequate to correct the unlawful condition; the negotiations break down; and the case goes to trial. As the statistics show, however, it has been possible in a great majority of such negotiations to arrive

[34] *United States* v. *General Electric Co.,* 82 F. Supp. 753, 798 (D.N.J. 1949).

[35] Henry Campbell Black, *Law Dictionary,* 533.

[36] *United States* v. *Radio Corporation of America,* 46 F. Supp. 654, 655 (D.C. Del., 1942).

at a settlement which achieves the relief that is necessary and is consistent with the objectives of the antitrust laws."[37] This assistant attorney general went on to say, "Over the entire history of antitrust enforcement programs approximately 25 per cent of our civil cases have been terminated by consent judgments. In recent years, the percentage has been substantially greater."[38]

Whether or not a defendant agrees to accept a certain decree rather than litigate depends upon such factors as expense of litigation, publicity, importance of patents—especially if compulsory licenses or dedication of them to the public is required—and, most important, possible outcome if a trial and a court decision, rather than a consent decree, is the chosen alternative. In connection with the last point—possible outcome—one should mention that a decree cannot be used as prima-facie proof in a triple damage suit while a decree or decision rendered after the introduction of evidence in court can be.

Whether a consent decree is lenient or strict hinges, it seems, upon the political and judicial climate; therefore its effect varies from one time to another. The antitrust division of the present may deplore the leniency of consent decrees in the past. Witness, for example, the efforts in 1942, as mentioned above, to vacate the RCA consent decree of 1932. In contrast, certain legal counsel today denounce present decrees and advocate those of another era.

Certainly, though, the consent decree is often used in resolving the legal complications frequently arising from and surrounding the cross-licensing of patents and the establishing of patent pools—as has often been done concerning a number of varied commercial products and items.

a. *Magnesium.* This metal, about one-third lighter than an equal volume of aluminum, is found extensively in earth and sea water in combination with other chemical elements. From 1927 to 1941, Dow Chemical Company was the only producer in the United States, and later remained the principal one. Dow and the Aluminum Company of America—Alcoa—had patents covering both production and fabrication of magnesium. According to a government complaint of 1942, Dow Chemical and a subsidiary of Alcoa entered into a cross-licensing agreement in 1927; Dow granted a preferential price to the subsid-

[37] MP Hearings, Part 3, 48.
[38] *Ibid.*, 47.

iary, which in turn discontinued the production of this metal; and in 1934 Dow joined an international pool. As a result of the 1942 government complaint, a consent decree cancelled the agreements which restricted the production and fabrication of magnesium; made the patents in question, but not the know-how, available to any applicant; made the production patents royalty-free during the war but subject to a payment of reasonable royalties thereafter; and made the fabrication patents royalty-free without restrictions.[39]

b. *Rail Joint Bars*. Rail joint bars, sometimes called splice bars or fishplates, are used to support and fasten together the ends of railroad rails. When these bars became worn, it was necessary, prior to 1925, to replace them with new bars. About this time the process of re-forming or reshaping worn bars was started, and it developed to such an extent that most railways by 1943 shipped such bars to mills for re-forming. The process is basically simple—a worn rail bar, after being heated, is placed in a die and subjected to heavy pressure, and thus formed into a bar with smooth surfaces. The cost of this re-forming is from one-fifth to one-third the price of new bars.

Three companies—Rail Joint, Woodings-Verona, and McKenna—each owned patents pertaining to the design of rail joint bars and also patents on the process of re-forming. The basic patent on this process was issued in 1910; but later many other patents were issued on numerous phases of the process. The validity of these patents was never passed upon by the courts. The three companies aided each other in securing such patents and exchanged information concerning them. Beginning about 1928, they pooled their patents through cross-licensing agreements and sublicensed seven other companies to use some or all of them. A government complaint of 1943 indicates that they divided the territory and customers among themselves and their licensees and fixed prices on both patented and unpatented designs of reshaped bars. A consent judgment of 1944 enjoined these objectionable practices and directed these companies to dedicate certain patents to the public.[40]

c. *Air-conditioning Equipment*. Air conditioning means the control

[39] *United States v. Aluminum Co. of America,* CCH Trade Regulation Reports, Supp. 1941–43 ¶ 52,776 (S.D.N.Y. Civil No. 18–31; Complaint and Consent Decree, April 15, 1942).

[40] *United States* v. *Rail Joint Co.,* CCH 1946–47 Trade Cases ¶ 57,469 (N.D. Ill. Civil No. 43–C–1295; Complaint, 1943; Final Judgment, 1944).

of the temperature and humidity of the air, either for personal comfort or industrial processes, by means of fans and humidifiers. The forerunners of modern air-conditioning equipment were developed just prior to World War I. The Auditorium Conditioning Corporation was organized as a patent-holding company by two leading manufacturers of such equipment. These two manufacturers and three other companies owned Auditorium's stock and assigned their patents to it under cross-licensing arrangements. Other companies became licensees upon payment of fees and royalties, and one of them also assigned an air-conditioning patent to it. In addition, Auditorium purchased numerous patents from others until it had at least forty by 1943. A government complaint of 1943 maintains that more than 90 per cent of all comfort and industrial air-conditioning installations were made by Auditorium licensees, all of which operated under restrictive conditions. A consent decree in 1945 dissolved the patent-holding company, prohibited price fixing and price discrimination, forbade suits for patent infringement or enforcement of license agreements, and dedicated the patents involved to the public.[41]

d. *Flat Glass*. The three outstanding groups of glass products are flat glass, glass containers, and electric bulbs; patents covering all three of them have been abused. There are several kinds of flat glass —plate, window, safety, and tempered. Plate glass, in contrast with window, is ground and polished. The safety variety is a glass sandwich—two or more pieces of flat glass and a layer of plastic material between them that tend to stick together when broken. The tempered or flexible kind, made by heating and then suddenly cooling the glass, can withstand rapid and extreme changes in temperature without breaking. Plate glass is used mostly in automobiles, mirrors, and store fronts; safety, in automotive vehicles and railroad cars; tempered, in windows for insulation and in ovens and refrigerators and other places of extreme temperatures. The main factor in the location of glass factories is the availability of sand and gas.

According to a government complaint, two glass companies—Libby-Owens-Ford and Pittsburgh Plate—became dominant, by 1935, over the entire flat glass industry. They accomplished this by purchasing plants and patents from others, by becoming the exclusive

[41] *United States* v. *Auditorium Conditioning Corp.*, CCH 1944–45 Trade Cases ¶ 57,428 (S.D.N.Y. Civil No. 22–200; Complaint, 1943; Final Judgment, 1945).

licensees of foreign companies, and by cross-licensing each other. They eliminated competition in acquiring patents and developing processes.

The government also charged these two—Libby and Pittsburgh—and other glass companies with dividing domestic markets—limiting the customers to whom each company could sell and limiting the types of flat glass each company could distribute; with fixing prices on both patented and unpatented glass; with exchanging information relating to costs, sales, customers, and conditions; and with establishing intercorporate, interdirectoral and financial relationships with each other. The glass companies quoted only f.o.b. factory prices with freight equalized to destination.

A court decree involving the Corning Glass Works in 1946 required this company to get rid of its stock of American Security, to abandon any restriction as a condition to the exchange of information regarding present and future patents, and to grant compulsory licenses at reasonable royalties under patents relating to flat glass. A final judgment in 1948 covering the other nine glass companies was in keeping with this decree. An additional and significant feature is that Libby-Owens-Ford was ordered to dedicate to the public its sixteen flat glass patents relating to a certain process, and it and Pittsburgh Plate Glass were ordered to grant nonexclusive royalty-free licenses under 159 patents and to provide for five years technical assistance and information at actual cost. It should be noted that these companies in 1948 accepted royalty-free licenses in a consent decree although the Supreme Court in the Hartford-Empire case of 1945 had disapproved such licenses.[42]

e. *Lecithin.* Lecithin is a natural organic substance found in many animal and vegetable products. Most of it is a by-product in the processing of soybean and corn oils. It is used in such foods as shortening, chocolates, and bakery products and also in textiles, paints, leather dressing, and gasoline. Lecithin itself is not a patented product. Numerous patents, however, have been issued pertaining to processes for producing and refining lecithin, to its uses in various products, and to mixtures of it with flour and salt.

[42] *United States* v. *Libby-Owens-Ford Glass Co., et al.,* CCH 1948–49 Trade Cases ¶ 62,323 N.D. Ohio Civil No. 5239; Complaint, 1945; Final Decree *re* Corning, 1946; Consent Judgment *re* others, 1948); and *Hartford-Empire Co.* v. *United States,* 323 U.S. 386, 413–20 (1945), discussed in Chapter Three.

Prior to 1934, German and Danish corporations exported and sold most of the lecithin used in the United States. About this time, however, two domestic companies—Glidden of Ohio, and Archer-Daniels-Midland of Minnesota—began to make lecithin in connection with the production of soybean oil. In 1935, they incorporated the American Lecithin Company: 60 per cent of its stock was divided equally between them; 30 per cent, between the foreign companies; and 10 per cent was given to its officers. Two of its directors were chosen by Glidden, two by Archer-Daniels-Midland, one by American Lecithin, and one by the foreign companies. Further, American Lecithin received by assignment twenty-eight patents from abroad and twelve other patents and twenty-eight patent applications from the two domestic companies and its own employees and sales representatives. According to government charges in 1946, American Lecithin issued exclusive licenses under its pooled patents and secret processes to the two original companies—Glidden and Archer-Daniels-Midland—for the manufacture of lecithin in the United States and Canada, sold their output in equal proportions and fixed the prices, and let them obtain lecithin for further manufacture at 70 per cent of the current market price. The government also claimed that the foreign corporations had exclusive rights outside the United States and Canada and received a proportion of American Lecithin's profits. Beginning about 1939, additional companies had begun to produce lecithin in substantial amounts, two of which refused to become licensees.

A consent judgment in 1947 dissolved the joint sales agency, terminated suits for infringement prior to 1947, and enjoined questionable practices. In addition it required American Lecithin to license its "use" patents royalty-free and its "process" patents at uniform, reasonable royalties, all licenses without restrictions.[43]

f. *High-Tension Cable and Accessories*. High-tension cable consists of conductors of electricity separated and insulated from an outer sheath by oil or gas under mechanical pressure. This arrangement prevents the deterioration of the conductor and the leakage of electricity. Accessories include joints, fittings, and other essential parts. Fluid-filled cables and accessories are used in industrialized areas, and by railroads and public utility companies, for transmitting elec-

[43] *United States* v. *American Lecithin Co.*, CCH 1946–47 Trade Cases ¶ 57,542 (N.D. Ohio Civil No. 24115; Complaint, 1946; Consent Judgment, 1947).

trical power at high voltages. The development of this industry has been accompanied by a vast accumulation of patents and technological information.

Domestic manufacturers and sellers of such cable and accessories include General Cable Corporation, General Electric, Okonite-Callender Cable Company, and Phelps Dodge Copper Products Corporation. Okonite-Callender was owned equally by Okonite of the United States and Callender of Great Britain. According to a government complaint of 1947, these four companies, beginning in 1930, pooled their present and prospective patents and technology by means of cross-licenses; any company utilizing one or more patents of another paid a royalty to the pool in accordance with a schedule; all four of the participating companies shared equally the expenses of litigation in upholding the validity of these cross-licensed patents; and they acquired pertinent patents, inventions, and technology from outsiders, and added them to their patent pool. This complaint charged the four companies with refusing to license other domestic manufacturers and adhering to uniform prices and terms of sale under the leadership of General Electric.

A consent judgment of 1948 cancelled restrictive agreements and ordered the compulsory licensing of 314 patents in the pool and also any future patents applied for or obtained within five years, at reasonable royalties.[44]

g. *Ophthalmic Frames and Other Optical Goods*. Ophthalmic goods include frames—rimmed, semi-rimless, and rimless—and lenses used in eyeglasses and spectacles. The lenses usually require grinding before being attached to frames. In most instances this work is done by the wholesaler rather than the retailer because his volume of business is large enough to justify the necessary skill, machinery, and stock.

Four optical companies—American, Bausch and Lomb, Shuron, and Continental—made and sold frames and lenses; other companies limited their business to frames. According to a government complaint of 1940, the four optical companies licensed each other and also other manufacturers; sold only to "approved" wholesalers and retailers; fixed the prices of manufacturer, wholesaler, and retailer

[44] *United States* v. *General Cable Corp.*, CCH 1948–49 Trade Cases ¶ 62,300 (S.D. N.Y. Civil No. 40-76; Complaint, 1947; Consent Judgment, 1948).

for patented and unpatented frames, nonpatented lenses, patented and unpatented extras for frames, and labor charges for maintaining lenses; and employed an elaborate system of policing. The "kick-backs" to eye doctors presumably were for the purpose of bringing about their co-operation. American Optical and Bausch and Lomb, because of their size and ownership of many wholesaler outlets, were leaders in making these arrangements. In 1940, they accounted for 63 per cent of the frames made and sold in the United States, and five other co-conspirators accounted for 23.

A consent judgment of 1948 involved thirteen manufacturers of eyeglass frames, six optical wholesalers and their trade association, and two patent holding companies. The judgment cancelled patent licenses and other agreements through which prices had been controlled, required that pertinent patents and trade-marks be made available to other manufacturers on reasonable terms, restricted the expansion of certain manufacturers, and prohibited the boycotting of wholesalers.[45] The unusual scope of this decree suggests the ramifications of the alleged restraints and the importance of the product to public health.

h. *Elastic-Top Hosiery and Machinery*. In 1949, the government charged the Davis Company and three other corporations with obtaining a monopoly on elastic-top hosiery and the machinery for making it by pooling competing patents in these fields. A consent judgment of 1952 forbade Davis, the patent holding company, to bring infringement suits under any machinery patent or to acquire in the future any interest in any patent, and required this company to grant licenses under all its hosiery patents upon nondiscriminatory terms and to be dissolved upon the expiration of the last patent owned by it.[46]

i. *Tractor Cabs*. Bearing Distributors Company and two other manufacturers of tractor cabs, in a cross-license agreement, settled a three-way interference, became licensees at certain rates of royalty under the patents awarded for priority of invention and without royalty under any subsequent patents, and refused licenses to others.

[45] *United States* v. *American Optical Co.*, CCH 1948–49 Trade Cases ¶ 62,308 (S.D. N.Y. Civil No. 10–391; Complaint, 1940; Final Consent Judgment, 1948).

[46] *United States* v. *The Davis Co.*, CCH 1952 Trade Cases ¶ 67,403 (S.D.N.Y. Civil No. 54–357; Complaint, 1949; Consent Judgment, 1952).

A final consent judgment in 1953 required them to grant licenses at reasonable royalties under certain patents owned and under any other tractor cab patents acquired within five years.[47]

Complaints. Recent complaints of the Department of Justice, beginning in 1949, include the following instances of alleged pooling or exchanging of patent rights contrary to law. E. I. Du Pont de Nemours and Company, by acquiring stock control of General Motors and United States Rubber Company, brought about the exchange of patents and technical information exclusively or preferentially among themselves.[48] Four companies pooled and purchased competing patents on peach-pitting machinery, issued restrictive licenses, acquired competitive machinery, and thus monopolized and restrained trade in the manufacture and use of such machinery.[49] Producers of slip covers for furniture exchanged patents relating to this product.[50] Two companies pooled their patents on the hydraulic oil well pump, and this pool in turn acquired other important patents in this field.[51] Three manufacturers of stamped metal wheels exchanged patents.[52] Several companies agreed to pool their present and future patents and grant licenses so as to allocate the manufacture and sale of machine tools among themselves according to specific fields.[53] Two manufacturers of induction hardened crankshafts pooled pertinent patents and know-how and allocated fields of commercial development.[54] Several manufacturers of railroad car couplers acquired and pooled patents and issued licenses only to themselves upon restrictive conditions as to prices, territories, and sales.[55]

[47] *United States* v. *Bearing Distributors Co.*, CCH 1952–53 Trade Cases ¶ 67,595 (W.D. Mo. Civil No. 6895; Complaint, 1951; Consent Decree, 1953).

[48] *United States* v. *E. I. du Pont de Nemours and Co., et al.*, CCH 1950–51 Trade Cases ¶ 62,557 and 62,633 (N.D. Ill. Civil No. 49C–1071; Complaint, 1949).

[49] *United States* v. *Food Machinery and Chemical Corp.* (N.D. Calif. Civil No. 29308–G; Complaint, 1949).

[50] *United States* v. *Krasnov* (E.D. Pa. Civil No. 11024; Complaint, 1950).

[51] *United States* v. *Kobe* (N.D. Okla. Civil No. 2865; Complaint, 1951).

[52] *United States* v. *Kelsey-Hayes Wheel Co.* (E.D. Mich. Civil No. 10–655; Complaint, 1951).

[53] *United States* v. *Associated Patents, Inc.* (E.D. Mich. Civil No. 10–664; Complaint, 1951).

[54] *United States* v. *Ohio Crankshaft Co.* (N.D. Ohio Civil No. 28299; Complaint, 1951).

[55] *United States* v. *National Malleable and Steel Castings Co.* (N.D. Ohio Civil No. 30281; Complaint, 1953).

Patent Pools without Unlawful Monopoly and Restraint of Trade

As previously noted, pooling or cross-licensing of patents without restrictions as to output and price are within the law. In fact, in interpreting pools involving the cracking of petroleum and electric fuse cutouts, in 1931 and 1948, the Supreme Court pointed out the legality and even suggested the desirability of proper arrangements which enable owners of rival or complementary patents to reconcile conflicts so that the inventions may be used. Two outstanding pools, one in automobiles and the other in airplanes, which have lessened conflicts in technology and remained within the law, will be presented.

Automobiles. Practically all automobile manufacturers paid royalties on the Selden patent until 1911, when a circuit court of appeals held that Ford Motor did not infringe it.[56] In 1915, the members of the Automobile Manufacturers Association—to which Ford Motor, however, has never belonged—sought to avoid patent disputes in the future by authorizing this association to grant free licenses to each other under their patents then owned and under their future patents for the next ten years. Although several companies had no patents at all when they became members, they received licenses nevertheless. This agreement included parts manufacturers if subsidiary to members but excluded them if independent. Also it did not cover patents solely adaptable to trucks and other commercial vehicles. Another exception was any invention deemed revolutionary; however, the only effort to put it into effect was disallowed both by the board of directors and the three arbitrators of the association. To avoid possible controversy over such an exception, the next agreement, from 1925 to 1929, was confined to patents owned in 1925—at the beginning of the period. The next three agreements, 1930–34, 1934–39, and 1940–46, covered only the patents owned January 1, 1930. The next one, 1947–51, included the patents owned January 1, 1947, but the agreement beginning in 1952 included only those owned January 1, 1940.[57] The agreement beginning in 1952, by excluding patents issued since 1940, is the most restricted in scope; it applies to patents none of which has more than five years to run. What accounts for this change

[56] *Columbia Motor Car Co.* v. *Duerr and Co.*, 184 Fed. 893 (2nd Cir. 1911).
[57] TNEC Hearings, Part 2, 292.

in policy? Does it reflect the conflict of interest between members with parts-manufacturing subsidiaries having important patents and members without? Do some of the members have important patents which they refuse to share without royalties?

The following table lists the agreements and the patents to which they apply:

Period of Agreements	*Application to Patents Owned by Members on*	*Number of Patents and Year*
1915–24	Jan. 1, 1915	547—1915
1925–29	Jan. 1, 1925	1066—1925
1930–34	Jan. 1, 1930	1687—1930
1935–39	Jan. 1, 1930	1285—1935
1940–46	Jan. 1, 1930	1058—1938
1947–51	Jan. 1, 1947	Unknown
1952–	Jan. 1, 1940	Unknown

Ford Motor and members of the association have exchanged licenses free of charge. Packard Motor was like Ford Motor in accepting licenses and paying royalties and in granting licenses but unlike in collecting royalties. Since the creation of the association there has been no patent suit between automobile manufacturers. Royalties have been paid only on a few hundred patents, amounting to about $100 in royalties per car, and they generally go to patent owners who are not automobile manufacturers. However, there has been considerable litigation in the automotive industry as a whole. In 1952, for example, Ford Motor, in keeping with a consent judgment involving its alleged infringement of certain tractor patents, paid $9,250,000 to their owner.[58]

Members of the association still invent and take out many patents to avoid payment of royalties to outsiders. Parts manufacturers and independent inventors compete with each other and with automobile manufacturers in developing, making, and selling inventions. They may license manufacturers but usually sell parts to them for initial installation and for replacement and sell various gadgets only to car owners via middlemen. Licenses to a member of the association are nonexclusive.

In general, there are many independent producers of each part.

[58] *Harry Ferguson* v. *Ford Motor Co.*, (S.D.N.Y. Civil No. 44–482; Complaint, 1948; Consent Judgment, 1952). Also see 77 F. Supp. 425 (S.D.N.Y. 1948).

However, by 1939 there were only three manufacturers of automobile wheels; at least fifteen others had been liquidated or purchased. The three cross-licensed each other on a nonexclusive basis without royalties and restrictions, but a license was not granted freely to a newcomer.[59]

Airplanes. In 1916, the United States appropriated $15,000,000 for the purchase of aircraft, but patent litigation among the fifteen manufacturers in the field held up production. Since some of them, Wright-Martin especially, insisted upon very large royalties, Congress passed a law and an appropriation for condemning and acquiring their patents. In response to this threat and in accordance with a plan approved by the National Advisory Committee, the fifteen companies organized the Manufacturers' Aircraft Association in July, 1917, for pooling patents on airplane structures, but not on engines or instruments. Each member agreed to pay an initiation fee of $1,000 and a fee of $200 for each airplane built, $135 of which went to Wright-Martin, $40 to Curtiss, and $25 to the association until each of these two companies should receive $2,000,000 or until their patents expired in October, 1933. Quarterly reports from each member disclosed its production. The members waived claims against each other prior to July, 1917.[60]

In 1928, with the advent of the low-priced commercial plane, the royalty was reduced to 2 per cent of the selling price minus the cost of the power plant and propellers, with a maximum of two hundred dollars. After 1933, in keeping with the original agreement, no royalties were required except for administrative expenses. Therefore, in 1933 royalties were reduced to one-fourth of 1 per cent and in 1935 to one-eighth of 1 per cent of the selling price. With the exception of royalties and other slight changes the agreement of 1917 still remains in effect. Most airplanes have been manufactured under its provisions.

In reporting patents acquired since 1928, a member of the association may claim extra compensation, "in which case a commercial arbitration is held to determine whether the invention covered by the patent is of such a nature as to justify a monetary award."[61] According to a statement of its president in 1935, "At the present time

59 TNEC Hearings, Part 3, 1053.
60 Pooling Hearings, Part 1, 817–18.
61 *Ibid.*, Part 3, 2977.

there are approximately 750 patents licensed under the agreement now being administered by the association. Since December 31, 1928, arbitrations have been held on 159 patents reported with claims for special compensation. Of that number, 87 patents have been refused extra compensation, and 51 patents have received cash awards."[62]

The government has held manufacturers, practically all of whom except Ford Motor have been members of the association, harmless for any infringement of patents that might result from the production of planes for its account—a provision which has permitted the manufacturers to ignore the patents of others. Under this arrangement they have had no incentive to negotiate with an independent inventor about becoming his licensees and paying royalties. In case of infringement the independent inventor's only recourse has been to sue the government in the Court of Claims—a procedure so time-consuming and expensive that justice is practically impossible for him.[63]

Unfavorable testimony about this patent pool has been presented before Congressional committees by various persons. Perhaps Brigadier General William Mitchell was the most severe critic. According to his sworn testimony in 1935, members of the Manufacturers' Aircraft Association menace "any independent manufacturer with the ever-present threat of suit for infringement of patents in case he attempts to construct airplanes. At the present time, it is a question whether many of their patents are worth anything. Although most of the patents held by the Manufacturers Aircraft Association are allegations, private manufacturers did not want to test them in the courts, as it requires so much money to fight the case. Consequently, competition between the best aeronautical minds in the country is diminished, minimum change takes place in the output, constant effort is made by the holders of these patents to maintain the pool and have the Government contract for the construction of airplanes by negotiation instead of open competition."[64] General Mitchell went on to say, "At present, this pool discourages initiative on the part of the independent builder, thus keeping down air development."[65] For illustration, General Mitchell pointed out that "Henry Ford had one of the best equipped airplane factories in this country and turned out ex-

62 *Ibid.*, 2977.
63 *Ibid.*, Part 1, 202; Part 3, 2972.
64 Pooling Hearings, Part 1, 6.
65 *Ibid.*, 819.

cellent equipment, but he would not join the Manufacturers' Aircraft Association and was harassed by them, whereupon, for this and other reasons, he closed his factory."[66] Furthermore, "These aircraft companies have very little that has not been developed, directly or indirectly, by the Government."[67] It was General Mitchell's opinion that the independent inventor did not enter the association and get the benefit of all its patents because he considered his patents too valuable to turn in and he did not have the thousand dollars for initiation fee.[68] According to his statement in 1935, "Over $1,500,000,000 was spent on aircraft during the war and only a few worthless aircraft reached the front with which to fight the enemy in France. . . . During the 17 years since the war, about an equal sum has been spent on aircraft, and I can tell you that today the United States military service has not one single airplane in service with the units which is suitable for engaging in war against a first-class power."[69]

In comparing the automobile and aircraft pools, Mitchell declared that "in the automobile industry their products are sold to the public and excellence of output must dominate, otherwise the public will not buy. In the aircraft industry, however, where about 70 or 80 per cent of the output is taken or paid for by the United States government, most of the remainder being sold to foreign countries (the development for which has been paid for by the U. S. government) conditions are somewhat different. Under such conditions, the manufacturer whose principal object is financial gain will usually subordinate continuous improvement, which necessitates maximum change and minimum profit, to maximum profits and minimum change."[70]

As already indicated, Mitchell contended that the manufacturers of airplanes who were members of this association and the Aeronautical Chamber of Commerce, did not compete, in either quality or price. His conclusion was that "aeronautical equipment should be purchased after a design competition, a flying competition, and a price competition."[71] He recommended also that Congress "do away with

[66] *Ibid.*, 824.
[67] *Ibid.*, 824.
[68] *Ibid.*, 863.
[69] *Ibid.*, 820.
[70] *Ibid.*, 818–19.
[71] *Ibid.*, 12.

the private pooling of patents, and in its place establish a system by which the Government will control the aircraft patents directly and be able to license whoever they desire under them."[72]

According to a 1951 statement of the association, technical progress has been made available to all manufacturers without litigation or monopoly. Membership "has never been restricted in any respect, no qualified applicant has ever been refused the right to acquire licenses under the terms of the agreement, and there have never been any withdrawals from the association except in the case of companies which have either gone out of business or have ceased the manufacture of aircraft. . . . There has been no price fixing, no regulations or control of markets, nor any other restriction in regard to the sale of products."[73]

During World War II the government "was informally licensed under a pool of more than 600 patents held by the Manufacturers' Aircraft Association, not being required to pay royalties when planes were manufactured either by the government or a member of the association."[74]

The airplane pool, like the automobile or any other, has avoided conflict of technology and therefore litigation among the members. This was especially important at its beginning in 1917. However, this very harmony—pooling all patents of members and giving each the right to use the inventions of the other—took away each member's incentive for basic inventions either through its own research or through purchase from an outsider. In fact, such a group generally opposes any innovation that would depreciate what they already have. Revolutionary changes in aviation have come from outside the pool—for example, the jet engine from an independent inventor in another country, and war planes from the research directed and financed mainly by the government. And, since the members agree to accept each other's patents without question, there has been practically no occasion to adjudicate their validity and therefore determine the extent to which they monopolize ideas which rightfully belong to the public.

Television. A committee of engineers from some thirty companies, under the leadership of the general manager of the electronic division

[72] *Ibid.*, 14.

[73] *Summary of Operations and Outline of Services*, 5.

[74] *Patents, Processes, Techniques and Inventions*, Surplus Property Administration *Report*, (1946), 11.

of General Electric, succeeded in finding a new signal or electrical impulse which simultaneously conveys pictures in color and in black and white and thus makes possible both color television and the continued use of "black-and-white" television sets. These thirty companies, in achieving their goal, pooled the discoveries and resources of their laboratories and thus ignored patent rights that might have existed.

Conclusion

Pooling or cross-licensing of patents between two or more rivals, each merely permitting the other to use its patents, is within the law. Their licensing of competitors on a royalty basis and without price and kindred restrictions, giving licensees the complete right to use their inventions, is legal; in fact, it may actually promote competition. Any royalty-sharing scheme, like that in the gasoline-cracking case, is permissible, provided licenses are available to all on a nondiscriminating basis and a low percentage of interstate commerce is involved. However, if former competitors pool their patents on the basis of restrictions as to prices, such an action is illegal. And if they license other competitors under such restrictions, the arrangement is likewise unlawful.

In the Line Material decision of 1948—as in the Hartford-Empire opinion of 1945, presented in the next chapter—the Supreme Court sanctioned compulsory licensing with reasonable royalties. However, a few companies accepted consent decrees requiring compulsory licensing without royalties, in spite of the previous refusal of the Supreme Court in the Hartford-Empire case to sanction such a measure.

CHAPTER III

Patent Consolidations

THE ILLEGALITY OF CERTAIN patent pools has encouraged the outright and single ownership by a manufacturer of all patent rights pertaining to a particular industry. In carrying out this scheme both plants and patents or only patents may be acquired. According to government reports and court decisions, any one of several objectives may be the incentive. By controlling alternate inventions and the improvements upon them, the monopoly of an industry may be brought about. Another purpose may be to avoid the scrapping of existing equipment. Also, the best ideas of various inventions may be combined into a superior product or process. Lastly, other patents may be obtained to avoid litigation, although in some cases alleged interference or infringement may serve only as an excuse.

A company may acquire patents from outsiders or from its own inventors, laborers, and licensees. In purchasing the patents of rival companies, a manufacturer may insist that the rivals remain out of business. In some companies, especially large ones, the most important source of patents is usually the corps of employed inventors. These men receive fixed salaries and assign their patents. They attack their problems and work them out in well-equipped laboratories. Other employees—technicians and patent attorneys—prosecute applications for patents in the most effective manner, making validity searches, and carrying through elaborate and expensive interference and infringement suits. The employer is ever seeking new patents to replace those about to expire.

Within recent years the cost of research of large corporations has been borne greatly by the government, particularly by the army and navy. During World War II the government gave almost one billion dollars for research to nearly two thousand industrial organizations, and the top ten corporations received two-fifths of the total. In over

90 per cent of these research contracts the resultant patents went to the contractor. A research program on this basis still continues.[1]

In addition, the large corporation frequently requires its laborers, those who attend machines and perform other kinds of skilled labor, to assign their inventions, and the corporation's agreements with licensees usually require them to transfer to the corporation any improvements upon the licensed invention. Since licensee-users may be the most important source of such improvements and also of competition, the patent-holding company either takes their improvements in this manner or refuses to grant licenses at all.

There are, of course, exceptions to this general policy of buying up patents. A corporation may refuse to purchase improvement patents of outsiders which will expire about the same time as the basic patent. Moreover, its own policy or its purchase agreement with so-called rivals may result in turning down questionable patents or offering prices lower than patent owners will accept.

At least three-fourths of existing patents are issued or assigned to corporations; from 1921 to 1938 the patents issued to 157 large corporations rose from 3 per cent to 17.2 per cent of the total, and the ratio of their patents to total assets increased from 6 per cent to 16 per cent.[2]

Details about the methods of acquiring patents, the reasons for obtaining them, and the sources of them, appear in the following historical account of various companies.

Indifference and Leniency, to 1940

The extent of antitrust prosecution, particularly of companies whose power rested largely upon patents, and the judicial interpretation of them, have changed noticeably. About 1940, indifference and leniency in general gave way to a more aggressive prosecution and court decisions and decrees which reflect, as never before, the purpose of the Sherman Act. This change acquires meaning as various consolidations over the years are described.

Wire. The American Steel and Wire Company of New Jersey was organized in 1899 with a capitalization of $90,000,000 to manufacture

[1] CSB Report, 9–10.

[2] Comp. L. Hearings, 10; TNEC Hearings, Part 3, 1125, 1126.

principally barbed wire, woven wire, plain wire, and wire nails. It furnishes the first important illustration of an industrial monopoly built up largely through the control of many patents. This monopoly controlled 85 per cent of the manufacture and sale of barbed wire in the United States. In justifying the high price of the monopolized commodity, the following testimony was given by John W. Gates, the chairman of the company:

> But our control has been acquired by patents, which we have either gone and bought in the open market or paid others for, and we have paid out—in purchases of patents and in the litigation to establish them in the last 20 years—I should suppose, $5,000,000 or $6,000,000. They have paid out millions of money, and if you would eliminate that, I presume that prices might have been lower. I am willing to concede that.[3]

By means of patents this company also acquired a monopoly of woven wire fencing as well as barbed wire. The existence of this monopoly was revealed in the following testimony:

> There is a great deal of woven wire fencing made in this country. We claim a monopoly under patents of most of the woven wire fencing. We made the past year about 25,000 miles of woven wire fencing under patents that we purchased, and we get a very much larger profit on that than we do on the plain fence wire or even barbed wire or wire nails, for the reason that we own the patents on the machines and patents on the product.[4]

He also testified that the elimination of much patent litigation was another objective.

Tobacco. The old American Tobacco Company, created in 1890, was a consolidation of five manufacturers of cigarettes which made 95 per cent of the total output of the United States. It owed its success, according to a government report, in no small degree to the control of important patents relating to the tobacco industry. In fact, the possibility of acquiring exclusive control of certain cigarette-making machines furnished a strong incentive to its creation. Its general policy was to obtain any and all patents relating to any phase of the tobacco business—removing tobacco stems, rendering tobacco harmless to the smoker, and making tobacco bags as well as manufacturing cigarettes and cigars. It dominated the cigarette business

[3] Industrial Commission *Report* (1900), I, 1010.
[4] *Ibid.*, 1033.

as the result of the special adaptability of cigarette making to the use of machinery and its control of cigarette machines. The importance of these factors was shown by the difficulties of American Tobacco in acquiring domination of the cigar business. In all but the cheaper grades of cigars the greater part of the process was carried on by hand, so that the small concern was at no disadvantage.[5] In 1911, the Supreme Court ordered the dissolution of this company but gave no specific attention to its patents.

Harvesting Machines. Cyrus H. McCormick, the founder of the harvester industry, had many competitors when the industry was in its infancy. But he purchased patents on improvements from others; his employees—professional inventors and manual laborers—conceived many inventions, the patents on which were assigned to McCormick's concern. The International Harvester Company, created in 1902, was an amalgamation of the five leading manufacturers of harvesting machines. However, the basic patents on nearly all the chief lines of harvesting machines—binders, reapers, mowers, rakes—had expired before its organization; the corn harvester, appearing in the nineties, was the only exception. Minor patents covering improvements on all these machines had not expired.

Cameras and Film. Eastman Kodak Company of New Jersey, organized in 1901 with a capital of $35,000,000, was owner of the capital stock of the Eastman Kodak Company of New York, a corporation engaged in the manufacture of photographic materials and supplies, and of two other corporations, one English and one Canadian, similarly engaged. Within about fifteen years the New York company acquired the patents and other assets of twenty competing concerns in the United States; the plants were dismantled and their business discontinued or transferred to the Eastman plants in Rochester. The officers and partners of these corporations and partnerships agreed not to engage in competing business for periods of from five to twenty years. Also Eastman Kodak obtained entire control in the United States of imported raw paper for photographic use. It acquired stock houses in the larger cities, which handled chiefly its own products.[6]

In 1915, a district court decided that "such acts, when taken together, are most significant and seem to me to indisputably disclose

[5] *Tobacco Industry,* Corporation Bureau *Report* (1909), Part 1, 31, 38, 63–67, 85.
[6] *United States* v. *Eastman Kodak Co.,* 226 Fed. 62, 63–64 (W.D.N.Y. 1915).

an intention to violate Section 2 of the Sherman Act."[7] A final decree by the same court in 1916, required the discontinuance of the monopoly and the establishment of competitive conditions, a plan for which the defendants were requested to submit to the court within a given period.[8] The district judge in Buffalo, where the case was first heard, immediately directed the dissolution of the company. The decree required Eastman Kodak within two years to sell its Premo and Century plants at Rochester, its Artura brand of photographic paper, and certain brands of photographic dry plates. The properties required to be sold were the principal ones acquired from other manufacturers of photographic goods.[9]

Telephone. American Telephone and Telegraph, one of the nation's largest business enterprises, derives its income mainly from dividends on the stock of its operating companies and from their so-called license payments for research, development, and other services. Through Western Electric and Bell Laboratories it controls the manufacture of equipment and research. It remained free from federal regulation until the creation of the Federal Communication Commission in 1934. Its rate of return on original investment has been very high.

American Telephone—sometimes called the Bell System—was founded upon the basic patents of Alexander Graham Bell. Prior to the expiration of these two patents in 1894 it adopted, according to a government report, the policy of acquiring improvement patents from Western Union and others for the purpose of maintaining its dominance in communication. Nevertheless, alternate inventions after 1894 resulted in appreciable competition in the exchange telephone field. In 1908, either because of public relations or because of fear of invalidity of its patents, it discontinued its infringement suits and permitted Western Electric to sell telephone-exchange equipment to local independent telephone companies but under restrictions that gave it a competitive advantage. At all times it has maintained a controlling patent position in long-distance telephone service.[10]

Under the cross-license agreement of 1920 with General Electric, extended in part to Westinghouse, RCA, and others in the radio

[7] *Ibid.*, 80.

[8] 230 Fed. 522 (W.D.N.Y. 1916); appeal dismissed, 255 U.S. 578 (1921).

[9] *United States* v. *Eastman Kodak Co.*, A–51, Final Decree, (W.D.N.Y. 1921).

[10] FCC Report, 214.

group, American Telephone asserted rights to manufacture, use, lease, and sell broadcast transmitting equipment and to furnish wire facilities for radio broadcasting, subject only to the radio group's right to erect and operate broadcast stations. These activities were contested by the radio group, and this led to new cross-licenses and other agreements in 1926. However, there was no free interchange of rights to use the patents in all fields. American Telephone retained exclusive rights under its patents in two-way telephone services, both wire and radio; while GE and RCA withheld similar privileges in their domains. Each one, in effect, kept and obtained the patent rights in its particular field. After selling its broadcasting business in 1926, American Telephone limited its activities in this field to furnishing wire facilities. By reason of its patent position, its extensive wires, and its policies, it has attained a virtual monopoly in furnishing transmission services for national radio programs. Prior to the Communications Act of 1934 it avoided any statement that its wire services to broadcasters was a part of its public service undertaking.[11]

This company enlarged its patent structure from the two original patents of Bell, to the ownership of 9,255 patents and 1,279 patent applications on January 1, 1935. Its purchases from outsiders, though only 3.5 per cent of this total, included the loading-coil patents of Pupin, the vacuum tube patents of De Forest, and the repeater circuit patents of Grissinger—the three most important groups of patents since its acquisition of the Bell patents. Moreover, it obtained license rights under an even greater number of patents through license agreements with others; Western Electric, for example, obtained an exclusive license on amplifiers for talking pictures. In securing them, it has at all times attempted to acquire exclusive rights to the telephone and other fields in which it is primarily interested. It licenses radio, telegraph, and other companies under its patents but always with restrictions that preserve its dominance in the telephone business. Since it also licenses and cross-licenses others to avoid litigation, the courts have tested the validity of only a few of its patents.[12]

American Telephone has acquired important patents in other fields of communication. Its ownership of important patents on the telegraph printer or teletypewriter, for example, gives substantial control

[11] *Ibid.*, 214.
[12] *Ibid.*, 234, 246.

over Western Union, its chief rival. It furnishes the wire facilities to which this machine is attached, and a subsidiary of American Telephone is the principal manufacturer of this machine.[13] Its patented technology establishes the metes and bounds of other kinds of communication. Control over the inventions of a would-be competitor is more effective than ownership of its stock.

By amassing thousands of patents on inventions in the whole field of communication, mostly the result of intensive research in its own laboratories, American Telephone dominates the telephone and also controls "the exploitation of potentially competitive and emerging forms of communication."[14] It thus excludes others from its field and avoids being excluded by them. Would-be rivals may enter and remain only as licensees under restricted conditions. It pre-empts for itself new frontiers of technology for exploitation in the future and, in the meantime, protects what it has already developed. It keeps itself in a commanding position for the exchange of patent rights. In short, it employs patents to maintain its dominance and leadership in communication.

Incandescent Lamps. General Electric, incorporated in 1892, acquired Edison Electric, including all the Edison patents on incandescent lamps, and the Thompson-Houston companies. "Since that time, through the purchase and consolidation of numerous companies, through the purchase of patents, and through its own research organization, it has acquired most of the important patents covering electric lamps, their parts, and machinery and processes for making them."[15] GE and Westinghouse, according to an exchange agreement, received all patents of American Telephone and RCA relating to the lamp industry, in return for their patent rights applicable to the telephone and radio. According to a court decision of 1949, GE, after exchanging patents on machinery for producing glass, confined itself to "the production and sale of incandescent electric lamps," and Corning Glass Works, "to glass bulbs, tubing and cane for electrical purposes."[16] Also, for many years GE provided practically the only market for patents on lamps. Another source of its thousands of patents was its licensees. It "followed the conscious policy of funnel-

[13] *Ibid.*, 234.
[14] *Ibid.*, 584.
[15] *Incandescent Electric Lamps,* Tariff Commission *Report No. 133* (1937), 36.
[16] *United States* v. *General Electric Co.*, 82 F. Supp. 753, 798 (D.N.J. 1949).

ing into its control all patents held by its licensees and touching any phase of the industry. GE, however, minimized its privileges in this respect, contending that it had a mere nonexclusive license. The license included the right to sublicense for the duration of the life of the patent. It was thus placed in the position of rendering impotent any possible advantage to which one of its licensees would normally be entitled as the owner of a patent."[17] After the expiration of the last of its basic patents on the incandescent lamp in 1933, "it marshalled an offensive of patents of lesser degree to maintain its paramount position of control in the industry."[18] Again, "it individually monopolizes patents employed in the incandescent electric lamp industry."[19] The terms of the final judgment, approved by the court and presented in the next chapter, included the dedication of its incandescent lamp patents to the public.

Radio. General Electric organized the Radio Corporation of America in October, 1919, and stockholders included GE, Westinghouse, United Fruit Company, and American Telephone and Telegraph. Under a license agreement of November, 1919, GE, in exchange for stock, granted RCA substantially exclusive radio communication rights under its present and future patents, and provision was made for the mutual exchange of information about radio. There was a similar arrangement between RCA and Westinghouse, Marconi's Wireless of Britain, International Radio, Radio Engineering, United Fruit, and American Telephone. These various companies had patents on different means for the reception and transmission of messages by radio. These patents, though overlapping and conflicting in some instances, were competing, and apparently the combination of them in RCA was primarily to monopolize and restrain trade. This consolidation of radio patents—some two thousand—kept everyone else out of the radio field.

The cross-licensing contracts, which resulted in the creation of RCA, apportioned the entire field of wireless among the participants and protected them from competition with each other. GE and Westinghouse divided the manufacture of radio equipment, 60 per cent and 40 per cent respectively, and RCA was the sole seller of it. The

[17] *Ibid.*, 815.
[18] *Ibid.*, 815.
[19] *Ibid.*, 817.

two electrical companies withdrew from the sale of broadcasting transmission equipment and left that field to American Telephone, which in turn eliminated itself from the manufacture and sale of radio receiving sets. United Fruit agreed not to compete with RCA in wireless communication, withdrew from the sale of radio apparatus by conveying to GE the stock in its manufacturing subsidiary, and agreed to purchase wireless and radio devices from RCA. The entire arrangement divided the telephone, electrical, and radio fields and established supremacy in each through exclusive licenses.

As the result of its monopolistic position, RCA was the only American radio company engaged in transmitting and receiving radio messages between the United States and foreign countries. It controlled all the high-power stations except those owned by the government. In addition it entered into traffic agreements with various foreign governments and radio companies, the majority of the agreements providing that all messages to the United States must be transmitted only through RCA facilities. Its refusal to sell or lease apparatus to competitors for international communication was a well-defined policy. It paid independent inventors and small concerns for rights under their patents.[20]

The United States brought action for violation of the Sherman and Clayton acts. A consent decree of November 21, 1932, required the two electrical companies—GE and Westinghouse—to dispose of their RCA stock by distributing it ratably to their common stockholders, and enjoined exclusive patent licenses and division of field or territory. Nevertheless, domination of the telephone, electrical, and radio industries by the same companies has continued. However, RCA, in licensing other manufacturers of radio sets has imposed no restrictions on quantity, price, or markets—a situation conducive to the rapid strides in radio which subsequently took place.[21]

As already noted, the Department of Justice in 1942 failed to vacate the consent decree of 1932. In November, 1954, it charged RCA with a policy of "package licensing" its ten thousand patents in the radio-television field, according to which manufacturers must buy rights under all of them rather than one or more as they would prefer.

[20] Pooling Hearings, Part 4, 3395–96.

[21] *United States* v. *Radio Corp. of America,* CCH 1932–37 Trade Regulation Reports, Ct. Dec. Supp. ¶ 7025 (D. Del. Eq. No. 793; Petition, 1930; Consent Decree, 1932).

Strict Interpretation after 1940

In the judicial renaissance, beginning in the forties, the concentration of patents played a part in the interpretation of violations of the Sherman Act. Courts looked with suspicion upon the practice, and the Supreme Court approved compulsory licenses with royalties. Consent decrees in several instances provided compulsory licenses without royalties and even the dedication of patents to the public. Some corporations, whose judicial history extends back several decades, have experienced both lenient and strict treatments.

Glass Containers. Glass containers provide an outstanding example of the Industrial Revolution in recent times. Until after the turn of the century they were made only by glass blowers. A skilled workman would shape a gob of glass on the end of a hollow tube, lower it into a mold, and then blow the gob until it conformed to the walls of the mold. The invention of power-driven machinery to perform these tasks revolutionized the industry. The results were cheaper and better products, a seven-fold expansion of output, fewer workers, larger factories, and fewer companies.

There are two fully automatic methods for making glassware—suction and gob-fed. In the suction type the molten glass is sucked up directly into the molds. In the gob-fed type the molten glass is dropped by a feeder into molds which are parts of a separate machine. After the glassware leaves the forming machines it is carried by a conveyor into a lehr or closed tunnel, and there it is stacked on a belt which moves slowly through an annealing chamber. The glassware is cooled gradually to avoid cracking.

The government, in a complaint in 1939, charged the manufacturers of glass containers with violation of the antitrust laws. The disclosures of the Temporary National Economic Committee were largely responsible for the institution of this proceeding. The combination was characterized by District Judge D. Frank Le Blond Kloeb as "an unholy alliance" and by Justice Hugo Black of the Supreme Court as a "successful economic tyranny." The complaint was concerned with the suction and gob-fed processes, forming machines, stackers, and lehrs. The most powerful companies in the trust were Owens-Illinois Glass, Corning Glass Works, and Hartford-

Empire. Others, of less importance, included Hazel-Atlas Glass, Thatcher Manufacturing, Lynch Corporation, and Ball Brothers.

M. J. Owens, a glass blower who invented the suction process, produced in 1903 the first fully automatic machines for blowing bottles. His company, beginning about 1904, granted exclusive rights to certain licensees in limited fields for the making of glassware and acquired substantial stock interests in the licensee companies. In 1909, the Owens Company began to manufacture bottles and gradually shifted its efforts away from the manufacture of machinery except for its own use. The Owens machine dominated the industry until 1916. This company and another were merged in 1929 and became Owens-Illinois—the largest manufacturer of glass containers.

Corning, a large manufacturer and distributor of incandescent bulbs, signal and optical ware, and heat-resisting articles, organized in 1909 the Empire Machine Company as a patent holding and developing corporation. Hartford-Fairmont Company, organized in 1912 by Beech Nut Packing and Monongah Glass companies and by a group of Hartford, Connecticut, engineers and inventors, acquired broad patents upon the gob machines. A clash of patent applications between these two groups—Empire Machine and Hartford-Fairmont—led to an agreement in 1916. Corning was granted exclusive license to use Hartford-Fairmont's patents for pressed and blown glassware and the latter in turn was given an exclusive license to use Corning's patents for manufacture of glass containers. This agreement, very satisfactory from a business standpoint, led to the formation of Hartford-Empire in 1922. This company received all the glass machinery patents of these two groups and thus extended its domination of gob-feeding patents in the glass container industry. Its income was derived chiefly from royalties and fees obtained from the licensing and leasing of its machinery, which was made by another company.

In the meantime, the Owens company, exposed to keen competition from the cheaper gob feeders, developed and acquired patent rights on gob feeders and threatened to fence in its adversary. Owens and Hartford-Empire clashed in the Patent Office and the courts, and, as a result, they entered into a cross-licensing agreement in 1924. According to its terms, Owens granted Hartford-Empire an exclusive license under its patents on feeder and forming machines and paid royalties except on forty machines. In return, Owens had the privilege for eight years of preventing Hartford-Empire from licensing any bottle manufacturer whom Owens found objectionable; and Owens received one-half—only one-third beginning in 1932—of Hartford-Empire's divisible income from licenses over and above $600,000. Though excluded from the pressed and blown field previously reserved to Corning, Owens kept full control of its patented suction process. It obtained an option on all Hartford-Empire inventions in this field, purchased suction rights from others, and developed its own inventions.

After making this agreement in 1924, Owens and Hartford-Empire set out to obtain control of all other feeder patents. In this program they pooled the efforts of their legal staffs and contributed equally to the purchase of patents and the expense of litigation. In 1925, for example, they purchased the feeder patents of the W. J. Miller Company for $145,000 and those of the Federal Company, with whom they had been in interference proceedings, for $1,600,000. Hazel-Atlas, second to Owens in the manufacture of glass bottles, offered strong opposition by using feeders of its own design and manufacture. It refused to become a partner-licensee and Hartford-Empire brought infringement suits. Hazel-Atlas' legal expenses from 1926 to 1932 ranged from $50,000 to $150,000 each year and, in 1932, after losing an infringement suit which might have necessitated the payment of heavy damages and after being tempted by an attractive offer, it decided to enter the fold; it licensed Hartford-Empire under all its patents on glass machinery and in turn became another licensee. The financial terms brought satisfactory results. Hazel-Atlas received $6,500,000 from 1932 to 1938 as its share of the patent profits; it made license and royalty payments of $5,750,000. As a result of this and other arrangements the opposition to Hartford-Empire collapsed and practically every rival took a license.

As already indicated, forming machines, stackers, and lehrs also are important in the manufacture of glassware. Hartford-Empire and Owens obtained patents on them and thus strengthened their control over automatic glass manufacture. Several forming machines not covered by their patents were on the market; however, Hartford-Empire became dominant in this field by requiring the use of its forming machines with its feeders and by entering into a restrictive agreement with Lynch, its principal competitor. Through litigation and purchase of patents, particularly from the Thatcher company, Hartford-Empire became dominant also in stackers and lehrs.

Thatcher Manufacturing, a large producer of milk bottles, obtained an exclusive license to manufacture them on the Owens suction machine soon after its invention. In 1920, it secured the exclusive rights to make them on Hartford-Fairmont's feeder and forming machine. It pressed for exclusive rights under a later improved device, the single feeder. Hartford-Empire refused, but from 1925 to 1936 allowed Thatcher a rebate; in 1936 it agreed that, so long as Thatcher made 750,000 gross per annum, it would grant no other license for manufacture of milk bottles. All other milk bottle manufacturers except Liberty Glass were gradually brought under the Hartford-Empire licensing system with restrictions on quantity. Owens-Illinois and Thatcher manufactured and sold about two-thirds of all milk bottles in the United States.

Ball Brothers, the largest manufacturer of domestic fruit jars, had used machines of its own design as well as the Owens suction machines under license. In 1933, Hartford-Empire granted to Ball a license and also its residual rights in fruit jars and obtained an option to take licenses on all Ball's patents for glass machinery then owned or thereafter acquired. As a part of this transaction, several companies, including Brockway, sold their fruit jar rights to Ball. By oral agreement, Hazel-Atlas was limited to an output of 300,000 gross of fruit jars per year and Owens-Illinois to 100,000. In 1939, the year the government filed its complaint, Ball had 54.5 per cent of the output in this country, Hazel-Atlas 17.6 per cent, Owens-Illinois 6.4 per cent, and Kerr Glass Company, an outsider using a machine on which the patents had expired 21.5 per cent.

Hartford-Empire licensed the use of its machines with restrictions on the kind, quantity, and market of glassware—even on the weight, capacity, color, and composition. It did not impose price restrictions in its licenses because they were unnecessary; a system of price leadership prevailed through other restrictions and retention of title to licensed machines. Owens-Illinois was the price leader in narrow neck glassware, Hazel-Atlas in wide mouth glassware, Ball in fruit jars, and Thatcher in milk bottles. Price cutting was used when necessary to bring a competitor into line or force him out of business.

As might be expected, the close control of the glass container industry resulted in high prices and attractive profits. From 1925 to 1929, in spite of technological advances, prices declined only 10 per cent; from 1929 to 1933, in spite of the depression, only another 10 per cent. From 1934 to 1937, Owens-Illinios and Hazel-Atlas had profits on net worth that averaged around 14 per cent, and other licensees did as well or better; during this same period, when income from license fees and royalties practically trebled, Hartford-Empire's rate of profit averaged still higher.

In 1942, Judge Kloeb held that the glass companies had violated the Sherman and Clayton acts.[22] In January, 1945, the Supreme Court, to which the case was appealed, summarized the situation brought about in the glass industry, and existing in 1938, as follows: "Hartford-Empire, with the technical and financial aid of others in the conspiracy, had acquired, by issue to it or assignment from the owners, more than six hundred patents. These, with over one hundred Corning controlled patents, over sixty Owens-Illinois patents, over seventy Hazel-Atlas patents, and some twelve Lynch patents, have been by cross-licensing agreements, merged into a pool which effectually controlled the industry. This control was exercised to allot production in Corning's field to Corning, and that in restricted classes within the general container field to Owens-Illinois, Hazel-Atlas, Thatcher, Ball, and such other smaller manufacturers as the group agreed should be licensed. The result was that 94 per cent of the glass containers manufactured in this country on feeders and formers were made on machinery licensed under the pooled patents."[23] It agreed with the lower court in stating that these companies, "over a period of years, regulated and suppressed competition in the use of glass-making machinery and employed their joint patent position to allocate fields of manufacture and to maintain prices of unpatented glassware."[24] Further, it held that Hartford-Empire and Lynch had contracted with each other in violation of Section 3 of the Clayton Act.[25]

The district court ordered, among other measures, unrestricted, compulsory, royalty-free licensing under existing patents and patent

22 *United States* v. *Hartford-Empire Co.,* 46 F. Supp. 541 (N.D. Ohio, 1942).

23 *Hartford-Empire Co.* v. *United States,* 323 U.S. 386, 400 (Jan., 1945); other facts were obtained from this decision and also from TNEC Hearings, Part 2, *Patents.*

24 *Ibid.,* 406–407.

25 *Ibid.,* 401.

applications of these companies. The Supreme Court, though holding this royalty-free requirement improper, decided that the glass companies should be compelled to license their present and future patents in certain fields at standard royalties without discrimination or restriction and that they could not bring suit for past infringement of their patents. The Court even dissolved the trade association since it might become again an important instrument of restraint.[26]

In April, 1945, the Supreme Court gave another opinion clarifying its previous decision, and the case then was remanded to the district court for determination of royalties and other terms.[27] The final settlement, in 1947, provided that Hartford-Empire, upon demand, must sell rather than lease its glass-making machines, and fixed the maximum royalties for patents. This final settlement of royalties, it is estimated, saved the licensees five million dollars annually.[28] Also this company was required to dedicate to the public, in 1950, three of its basic patents on glass feeders because they were fraudulently procured by false affidavits and other underhanded practices. Separate settlements with Owens-Illinois, Hazel-Atlas, Ball, and Lynch made glass-making machine patents available to the public at royalty rates to encourage their use.

The agreement of 1916 between Empire and Hartford-Fairmont, according to the government, was their first violation of the antitrust laws. In 1922, Hartford-Empire got under way. In 1939, the government filed its complaint against this company; the final settlement came in 1947. The eight years, 1939–47, seem long, but they are short compared with the twenty-three years, 1916–39, it took the government to start action.

Both decisions of the Supreme Court in this case were four to two, three justices not participating. Justice Owen J. Roberts delivered the majority opinion and Justices Harlan F. Stone, Stanley Reed, and Felix Frankfurter concurred. Justices Black and Wiley Rutledge agreed "with the Court's judgment insofar as it sustains the decree of the District Judge."[29] They supported with minor exceptions the remedies prescribed by the lower court. Justices Douglas, Frank Mur-

[26] *Ibid.*, 413–20, 428.

[27] *Ibid.*, 324 U.S. 470 (April, 1945).

[28] Sigmund Timberg, "Equitable Relief under the Sherman Act," *University of Illinois Law Forum* (1950), 638, n. 43.

[29] 323 U.S. 386, 435 (Jan., 1945).

phy, and Robert H. Jackson did not participate in the case. As already suggested in the previous chapter, these "majority" and "dissenting" opinions furnish a significant background for comparison with the consent decrees of later years.

Shoe Machinery. In 1899, United Shoe Machinery Company of New Jersey acquired the patent rights and other assets of five companies, dominant in the manufacture and sale of welt-sewing, outsole-stitching, lasting, metallic-fastening, and heel-attaching machines. Later it acquired and developed other types of machinery until it had a full line of major machines, except for upper stitching, and all types of minor machines of importance. Its principal competitor in recent years, Compo Shoe Machinery Company, confined its operations to cement sole-attaching machines. All its other competitors made only a single type or a few types of shoe machines. Over 80 per cent of its machines in shoe factories were there on lease rather than from outright sale. It reserved the right to add new parts to them and thus extend its patent protection. This company required that it service and repair all its machines on lease. It made parts and distributed them from depots in shoe manufacturing centers. Likewise it sold such things as cutters and irons, nails and tacks, eyelets, and wire—$25,000,000 worth in 1946. In these supply fields, according to a district court in 1953, "United has control of the market as is shown by the fact that it is supplying much more than half the demand. This control comes principally from United's power over the shoe machinery market."[30]

Next in order is a brief record of certain acquisitions. Between 1899 and the first government action against it in 1911, this company acquired the shoe machinery business and assets of fifty-seven individuals, partnerships, and corporations. Its most important purchase was the complete line of shoe machines of Thomas G. Plant, including 633 patents and patent applications in the United States and other countries. In 1923, it purchased the patents of General Shoe Machinery and employed the principal inventor; General Shoe, organized in 1917 by International Shoe Company and other shoe manufacturers, had machines ready for commercial exploitation. In 1924,

[30] *United States* v. *United Shoe Machinery Corp.*, 110 F. Supp. 295, 346 (D. Mass. 1953).

United bought A. S. Little's interest in certain patented processes and machines for lasting and outsole stitching. In several instances United acquired the patents of inventors and then employed them; for example, in 1936 and 1937 it bought the patent rights to two heeling machines and then hired both of the inventors.

Between 1918 and 1938, according to a district court in 1953, United spent roughly $3,500,000 to purchase inventions and machines. "Most of these were from moribund companies, though this was not true of the acquisitions underlying the significant Littleway process and the less significant heel-seat-fitting machines and patents, each of which was from an active enterprise and might have served as a nucleus of important, though, at least initially, not extensive competition."[31] From 1938 to 1953, United did not acquire any significant patents, inventions, machines, or businesses from any outside source. Although its fundamental patents had expired by 1953, its "aggregation of patents does to some extent block potential competition."[32]

Furthermore, "About 5 per cent of United's patents were acquired. United contends that the patents acquired by United have been generally to enable it to pursue development which might otherwise have been blocked, or to resolve patent controversies, or for hedging purposes. But most of these purposes could have been served by nonexclusive licenses. Taking the further step of acquiring the patents has buttressed United's market power. In some instances . . . the acquisitions made it less likely that United would have competition."[33]

According to the courts, this company contracted with 95 per cent of the inventors of shoe machinery for the entire output of their inventive skill. By controlling the fountain of inventive genius it perpetuated its dominance. Inventors assigned their inventions to it not only during their terms of employment but also for stipulated periods thereafter. In 1947, it and its affiliates had 3,915 patents. Its research division employed about 575 men, and its patent department at least twenty-five lawyers who worked in close collaboration with those who were attempting to contrive new inventions or make improvements on the old.[34]

[31] *Ibid.*, 339–40.
[32] *Ibid.*, 339.
[33] *Ibid.*, 333.
[34] *Ibid.*, 330, 332.

The percentages of important types of machines which United had out in 1911 and 1947 are as follows:[35]

Machines	*1911*	*1946*
Clicking	100%	97%
Pulling Over	100	99
Lasting	99.9	94
Welt-sewing	100	97
Outsole-stitching	100	92
Loose-nailing	99.8	98
Heeling	94.7	93
Slugging	98.7	91
McKay Sewing	99.9	88

For many years United made a full line of shoe machinery; but in 1920 it apparently conceded the manufacture and sale of upper fitting machinery to Singer. Aside from this, its chief competition was the cement sole-attaching machine of the Compo company; in this field United's percentage of all machines was only 40 per cent. In 1947, most major operations in manufacturing shoes were performed by United machines. Its monopoly of certain basic machines through patents provided the leaverage for extending and maintaining control over other machines, supplies, and parts.[36]

The judicial history of United has been extensive. The Supreme Court, in 1913, upheld its legality as a consolidation of several shoe machinery concerns. "On the face of it the combination was simply an effort after greater efficiency. The business of the several groups that combined, as it existed before the combination, is assumed to have been legal. The machines are patented, making them is a monopoly in any case, the exclusion of competitors from the use of them is of the very essence of the right conferred by the patents, and it may be assumed that the success of the several groups was due to their patents having been the best. As . . . they did not compete with one another, it is hard to see why the collective business should be any worse than its component parts."[37] Such a decision is very questionable from an economic standpoint. It permits patents, regardless of

[35] *Ibid.*, 304–305; Brief for the United States No. 207, p. 150.
[36] *Ibid.*, 110 F. Supp. 295, 303, 339 (D. Mass. 1953).
[37] *United States* v. *Winslow*, 227 U.S. 202, 217 (1913).

number and purpose of acquisition, to give immunity, and it accepts vertical integration without question.

The Supreme Court in May, 1918, by a vote of four to three, announced again the legality of this combination of manufacturers of shoe machinery. Its line of argument was like that in the previous case. It declared that acquisition prior to the government suit in 1911 gave a false impression by their number. "They added nothing of obnoxious power to the United company nor in any practical or large sense removed competition."[38] It pointed out that some of the acquisitions arose from a desire to end patent litigation. Justice John H. Clarke gave a strong dissenting opinion, with which Justices William R. Day and Mahlon Pitney concurred. Justice Clarke declared that competitive concerns had been acquired and that patents could not excuse the consequent monopoly. If the two judges who did not participate—James C. McReynolds and Louis Brandeis—had held with the dissenters, a five to four decision would have resulted the other way.

As stated in Chapter Six, the Supreme Court in 1922 declared unlawful the tying clauses of United, but because the clauses violated the Clayton Act rather than the Sherman Act.[39]

In 1947, the government took further action. A district court, whose findings have been noted already, held that United had violated antitrust legislation. It prescribed the following remedies: "First, that United must offer for sale every type of machine it leases, though it remains free to continue leasing if the leases are not for more than five years, are terminable short of that time upon payment of nondiscriminatory, reasonable rates, have no full capacity clause, and segregate charges for machines from charges for repair services; second, that United must discontinue acting as a distributor of other companies' supplies and must dispose of its branches and subsidiaries which manufacture nails, tacks, and eyelets; and third, that United must make its patents available upon a reasonable royalty basis to those wishing to manufacture shoe machinery and supplies."[40] An appeal was taken to the Supreme Court and on May

38 *United States* v. *United Shoe Machinery Corp.*, 247 U.S. 32, 54 (1918).

39 258 U.S. 451 (1922).

40 *United States* v. *United Shoe Machinery Corp.*, 110 F. Supp. 295, 297 (D. Mass. 1953).

17, 1954, by *per curiam* opinion—by the court as a whole—the judgment was affirmed.

The litigation against United Shoe Machinery, from its beginning to the present, is a dramatic picture of the conscious use of patents to avoid competition and a striking illustration of administrative and judicial futility.

Aluminum. The Pittsburgh Reduction Company, organized in 1888, was succeeded in 1907 by Alcoa—the Aluminum Company of America. From 1889 to 1909 these two companies, one after the other, had a monopoly, presumably legal, of the manufacture of aluminum. In 1889, Pittsburgh Reduction obtained by assignment Hall's patent on a practicable process for eliminating oxygen from aluminum oxide and, in 1892, became the exclusive licensee under Bradley's patent on an economical process of smelting without the use of external heat. After 1909, Alcoa acquired many patents on certain processes, and of course still had technological secrets. A thorough investigation of Alcoa by the Federal Trade Commission was completed in 1924, but the Department of Justice took no action until 1937. In 1942, a district court in New York decided the case for Alcoa.[41] The Supreme Court could not review it because a majority of the court were disqualified for one reason or another, and consequently Congress passed a special law to permit its review by a circuit court of appeals. This appellate tribunal referred the case to the second circuit and, in 1945, Judge Learned Hand held that Alcoa had monopolized the aluminum sheet metal market.[42] But this industrial giant has not been broken up. Has the government just won a law suit but lost a cause? Two other aluminum companies—Reynolds and Kaiser—emerged after World War II not because of this litigation, but because of government policy in the disposal of surplus property built or acquired during the war.

In 1950, a district judge, who refused to divest Alcoa of any of its property, was aware that, "as of September 30, 1949, Alcoa owned 775 patents; was exclusive licensee under 57 additional patents, and was nonexclusive licensee of more than 125 others."[43] Nevertheless he concluded "that Alcoa's dominating position in the patent field

[41] *United States* v. *Aluminum Co. of America,* 44 F. Supp. 97 (S.D.N.Y. 1941).
[42] *Ibid.,* 148 F. 2d 416 (2nd Cir. 1945).
[43] *Ibid.,* 91 F. Supp. 333, 386 (S.D.N.Y. 1950).

does not, at present, constitute a serious threat to effective competition."[44]

Parking Meters. Vehicular Parking, solely a patent holding company, acquired patents of questionable validity from various sources. The promoter wrote to two of the defendants, "The parking-meter business is now [in existence] and seems to offer ample opportunity for profit. From my study of the patent aspects it seems that very little has been done to create a patent monopoly. . . . The actual strength of the patent monopoly must be gotten from a number of patents covering individual types of parking meters. If the Doyle patents were pooled with the other patents owned by Park-Rite Corporation and the several patents and patent applications owned by Dual Parking Meter Company, I feel that this pooling would result in a first-rate step toward establishing a dominant position in the parking-meter field."[45] The obvious purpose was to close all roads leading to a particular industry.

Chemical Rust-proofing and Priming Materials. Chemical rust-proofing and priming means the application to a metal surface of materials which interact chemically with the metal surface so as to retard rust. The Parker Rust-Proof Company, organized in 1915 to exploit two United States patents purchased from British owners, became the largest manufacturer and distributor of such materials. Parker maintained a staff of chemists and technical experts to improve its products and processes and to assist its customers in carrying out the patented process. Parker acquired patents as the result of the work done by its own staff, and it also purchased important patents covering both materials and methods of application from patent-owners outside the company. In 1938, Parker's president testified concerning the need of a "patent structure" or "more patents, more patents, and more patents."[46] Manufacturers, who accounted for 97 per cent of Parker's business, were required to use the patented processes only with Parker materials. At all times, however, Parker had competition from other forms of corrosion prevention.

The principal officers and research assistants of American Chemical Patent Company, upon the sale in 1940 of its patents and physical

[44] *Ibid.*, 392.

[45] *United States* v. *Vehicular Parking*, 54 F. Supp. 828, 831 (D. Del. 1944); CCH 1944–45 Trade Cases ¶ 57,226.

[46] Comp. L. Hearings, 334.

assets to Parker, agreed also to assign any relevant inventions that American Chemical Patent might acquire in the future. Patent litigation then pending was settled by agreement so that American Chemical was permanently enjoined from carrying on its rust-proofing process. This agreement was declared unlawful. A district court ruled that its purpose was to eliminate American Chemical as a competitor in the rust-proofing field and to terminate its activities in research to improve such processes. The court enjoined Parker from enforcing any of the rights secured under the patents covered by that agreement.[47]

In 1940, after litigation and interference proceedings with the Curtin-Howe Corporation, Parker purchased the patent rights of that corporation for $225,000. Its acquisition included the product and process patents—and patent applications—of this corporation, any improvements upon them, any additional inventions prior to 1950, and any inventions of a research chemist and official of this corporation prior to 1943. The court justified this acquisition on the ground that Curtin-Howe never offered any substantial competition and that Parker merely exercised its legitimate rights to purchase. The importance of prospective competition or of greater concentration of patent power apparently received no consideration.[48]

Concrete-Block Machinery. Besser Manufacturing Company became dominant in concrete-block machinery by acquiring the patents in this field. In 1942, Besser and its chief competitor, Stearns Manufacturing Company, made an agreement with two inventors which enabled all of them "to live in peace and harmony. Stearns and Besser were each given the right to use the Gelbman and Andrus patents—but exclusively to them—and it was agreed that they were to have the benefit of, and right to use, any and all future improvements and patents covering concrete-block-making machines that might come from the fertile minds of either Andrus or Gelbman or both."[49] In 1946, Besser "purchased Flam's exclusive licenses, had Flam agree to stay out of the concrete-block machine-making business for eight years and to make sure that he did, it gave Flam a job for five years. Then it went further—it got exclusive rights to any future patents

47 *United States* v. *Parker Rust-Proof Co.,* 61 F. Supp. 805, 811 (E.D. Mich. 1945).
48 *Ibid.,* 811–12; also CCH 1946–47 Trade Cases ¶ 57,431.
49 *United States* v. *Besser Mfg. Co.,* 96 F. Supp. 304, 310 (E.D. Mich. 1951).

Flam might get during the five years he was working for Besser."[50] When the government brought action against Besser, the court required Besser to dispose of its holdings in Stearns, to grant licenses under its patents at reasonable royalties, and to give the lessees of its machines the option to terminate or continue their lease or purchase the machines.[51] The Supreme Court approved these remedies. "Compulsory patent licensing is a well-recognized remedy where patent abuses are proved in antitrust actions and it is required for effective relief."[52]

Cellophane. Du Pont's monopoly of cellophane rested largely upon patents and trade secrets, obtained at first from abroad and later largely from its own research. Also, according to a government complaint, it had exclusive agreements with the largest cellophane converters. In 1953, a district court in Delaware held that Du Pont had not violated the Sherman Act—in fact, that it had actually increased competition by developing and selling a new packaging material. However, this decision, like the one favorable to Alcoa in 1942, may eventually be overthrown by a higher court.[53]

Atomic Energy. The previous instances of consolidation reflect the initiative and monopoly of private corporations. To a slight extent there is a similar concentration of patents in the government, particularly in the Atomic Energy Commission. The first chapter has already indicated the exclusive rights of this commission in the field of fissionable material and the use of atomic energy in a military weapon. However desirable this arrangement may be from the standpoint of defense, there may be objections to it as there are to any other monopolistic consolidation of patents: danger of arbitrary exercise of power and the lack of rivalry among inventors resulting in a lack of alternate inventions and technological progress. It should be added that by 1954 the AEC, though still owning the atomic patents important for defense, had released several hundred patents for licensing. Some government departments, particularly the Department of Agriculture and the Department of the Interior, dedicate their patents to the public.

Other Products. Many other well-known companies, as shown by

[50] *Ibid.,* 310.

[51] *Ibid.,* 314; also CCH 1950–51 Trade Cases ¶ 62,773.

[52] *Besser Mfg. Co.* v. *United States,* 343 U.S. 444, 447 (1952).

[53] *United States* v. *E. I. Du Pont de Nemours and Co.,* 118 F. Supp. 41 (D. Del. 1953).

government complaints and consent decrees, consolidated and strengthened their position in certain fields by means of patents.

a. *Ventilating Air Filters.* Air filters include devices for cleaning the air in ventilating and air-conditioning systems. Charges in a government complaint indicate that the control of these products was brought about by a patent consolidation. In 1925, two companies created the American Air Filter Company, assigning it their patents and receiving licenses in return. In 1929, the capital stock of the two companies was acquired by American Air Filter, and their principal stockholders agreed not to engage in the air-filter business for ten years and to assign to American Filter any of their future inventions in this field. Four other companies were acquired with similar restrictions upon stockholders. At least six different inventors assigned their patents to American Air Filter, and four of them gave it the exclusive right to acquire their future inventions. Two different patent owners, one of them International Harvester, granted it an exclusive license under certain patents. In 1939, according to agreement, American Air Filter hired three former employees of a rival company and acquired the interest of two of them in a pending patent application; all three of them agreed not to engage in the air-filter business for a certain time after their employment. American Air Filter granted other companies, particularly after unsuccessful attempts to purchase them, licenses with restrictions as to types of filter, price, territory, and purchase of mesh and other supplies. It bought all its requirements of cellulose material for air filters from a certain company which in turn sold such material only to it. In 1940, American Air Filter began to make wire filter, the patent on which was owned by another company, upon the condition that the purchase and sale of wire by the two companies be limited to each other. American Air Filter also got rid of other companies as either actual or alleged infringers; for example, one rival agreed to a consent decree and liquidation, a second to withdraw from all air-filter activities until 1947, and a third to accept a license under restrictive conditions.

A consent decree of September, 1946, required American Air Filter and another company to dedicate fifty-one air-filter patents to the public and to license thirty additional patents at reasonable royalties. Further, they were prohibited from enforcing contracts to prevent the competition of others in the manufacture and sale of air-filter materials, from requiring purchase from themselves of unpatented supplies for air filters through threats of patent infringement suits, and from conditioning patent licenses in any other way to control unpatented products.[54]

b. *Cash Registers.* National Cash Register, organized in 1906, employed unfair practices which caused its competitors either to quit or to sell their business to it. A consent decree of 1916,[55] which enjoined certain tactics, was modified to permit National's acquisition of other companies—Ellis Adding-Typewriter in 1929, Remington Cash Register in 1931, and Allen-Wales Adding Machine in 1943. In 1944, the Supreme Court, five to three, upheld a district court in denying the right of the Allen-Wales company to intervene.[56]

Later National Cash Register, flush with victory, acquired the business and patents of the General Register Corporation of Pennsylvania, a subsidiary of the Bell Punch Company, Ltd. The transaction was effected through the National Cash Register Ltd., apparently for the purpose of concealing violation of the decree of 1916. In 1947, the government filed a petition to ascertain why National Cash Register and three of its officers should not be found in criminal and civil contempt for violation of this decree. The company was fined but the officers were not. A consent judgment of the same date enjoined National Cash Register from making agreements with General Register concerning the manufacture or distribution of any register and forbade National either to restrict any person in the use or sale of or to collect any royalties on certain specified patents.[57]

c. *Public Fire-Alarm Equipment.* The apparatus in the fire-alarm field includes alarm boxes, circuits, and central office equipment. The Gamewell Company, located in Massachusetts, long manufactured and sold over 90 per cent of the fire-alarm systems operated by cities and other public authorities. According to a government complaint,

[54] *United States* v. *American Air Filter Co.,* CCH 1946–47 Trade Cases ¶ 57,492 (W.D. Ky. Civil No. 574; Complaint, 1943; Consent Judgment, 1946).

[55] *United States* v. *National Cash Register Co.,* CCH Trade Regulation Reports, Ct. Dec. Supp. III ¶ 3130 (C.C.S.D. Ohio Eq. No. 6802; Petition, 1911; Consent Decree, 1916).

[56] *Allen Calculators* v. *National Cash Register Co.,* 322 U. S. 137 (1944); also CCH, F.A.L. Case No. 100 and 112.

[57] *United States* v. *National Cash Register Co.,* (S.D. Ohio, Cr. No. 7092; petition and Consent Judgment, Jan. 8, 1947); also CCH, F.A.L. Case No. 884.

Gamewell's dominance in this field, beginning about 1896, was attained by buying competitors, acquiring patents from outsiders, cutting prices, rigging specifications to prevent bidding by others, and threatening patent and other litigation. In 1946, fifty years later, the government brought action; a consent decree of 1948, in addition to requiring the licensing of pertinent patents on a reasonable royalty basis and dedicating certain trade-marks to the public, sought to prevent "closed specifications" and prohibited Gamewell from making any offer to furnish engineering services without or below cost to prospective public purchasers of fire-alarm equipment.[58] In 1951, a district court held that Gamewell had violated this particular provision and imposed corrective measures.[59]

d. *Stencil Duplicating Machines, Stencils, Ink, and Kindred Supplies.* A stencil is a sheet of strong, porous, long-fiber tissue coated with chemicals to render it impervious to ink. Letters may be cut upon it by removing portions of the coating with a stylus or typewriter, thus exposing the porous tissue and permitting ink to pass through it to form like letters upon a sheet of blank paper placed in momentary contact with it. Various types of stencils have been invented and used: prior to 1886, sheets of paper coated with varnish; about 1888, Brodrick's porous tissue base with a pliable wax coating; in 1912, Fuller's wet protein with a gelatin-compound coating; in 1924, Hill's cellulose stencil coated with a nitro-cellulose compound; and in 1928, a dry-protein stencil. Fuller's stencil of 1912 could produce thousands of copies, but frequently caused damage to typewriter keys. The cellulose stencil has been the principal type used in recent years. Until about 1933 the porous base, universally used, was a long-fiber tissue paper made by hand in Japan and called Yoshino. Beginning about 1941, Yoshino was supplanted in the United States by a similar tissue made by machinery. Ink, repairs, and replacement parts complete the list of essentials in stencil duplication. (It should be added that duplicating machines other than stencils include those that use movable type.)

Organized in 1884, A. B. Dick Company of Chicago became by 1899 the largest manufacturer and distributor of stencils and stencil

58 *United States* v. *Gamewell Co.*, CCH 1948–49 Trade Cases ¶ 62,236 (D. Mass. Civil No. 6150; Complaint, 1946; Consent Judgment, 1948).

59 *Ibid.*, 95 F. Supp. 9 (D. Mass. 1951).

duplicating machines in the United States. It also made and sold parts, ink, and other supplementary supplies, and repaired the machines. After 1921, its sales revenue—about ten million dollars per year—from these products averaged over 80 per cent of the entire revenue resulting from stencil and stencil duplicating purchases in this country. A. B. Dick's dictation of supplementary supplies and its participation in a cartel are described in later chapters.

The history of this company reveals its acquisition of important patents relating to its products. An agreement in 1912 provided for the assignment to Dick of certain patent applications which were in interference with its Fuller protein-stencil patent. Even by 1916, the Dick company had 128 patents. They were originally issued to thirty different inventors—sixty-seven of them to Dick and the remainder to other individuals—and they covered alternate methods rather than just one way of accomplishing a particular result. This fact points to a policy of obtaining any and all patents pertaining to duplicating machines and supplies in order to secure a monopoly.

According to a complaint of 1946, exclusive licenses from two companies—Dexter and Manning—under their patents for making tissue to supplant Yoshino, enabled Dick to keep other manufacturers from obtaining such tissue from about 1934 to 1942.

In 1948, consent judgments enjoined the questionable practices already described. In addition, the Dick company, though permitted to sell direct to a limited number of customers on an annual basis, was required to discontinue the operation of its retail sales branches for five years. Also, for fifteen years it had to refrain from acquiring any interest in any concern engaged in the manufacture or resale of stencil duplicating machines, stencils, and kindred supplies. And fulfilling the decree, it dedicated thirty-five patents to the public and disclosed the know-how for the commercial practice of them, along with its secret "web manufacturing" process for making stencils; discontinued collection of judgments from its patent infringement suits; and dedicated its trade-mark, "mimeograph," to the public. In the same judgments, the Dexter and Manning companies were enjoined from enforcing significant patents relating to the manufacture of stencils.[60]

[60] *United States* v. *A. B. Dick Co.*, CCH 1948–49 Trade Cases ¶ 62,233 (N.D. Ohio Civil No. 24,188; Complaint, 1946; Consent Judgment, 1948).

e. *Braking Systems.* In a braking system, the actuating equipment may be mechanical, hydraulic, or air. There was a shift, mostly during the thirties, from mechanical to hydraulic brakes for automotive passenger vehicles. Large busses and trucks utilize air or power braking systems. The manufacture and sale of braking systems was carried on by seven companies—Bendix Aviation, General Motors, Hydraulic Brake, Wagner Electric, Du Pont, Bendix-Westinghouse, and Westinghouse—and the allegations of a 1947 government complaint indicate the concentration of patents in this field.

By 1929, Bendix owned or controlled almost every important patent relating to mechanical braking systems. In its reorganization that year it gave 20 per cent of its stock to General Motors in consideration for the braking-system patents and some other assets. Afterwards, GM discontinued its sales of mechanical brakes to other automotive manufacturers.

Hydraulic Brake, organized in 1918 and owned by Bendix, sought to develop hydraulic braking systems; by 1924, it claimed a monopoly of patents on the only commercially feasible brake of this type; and, after 1924, it acquired hundreds of patents in this field. This company issued licenses to install its hydraulic braking systems, on a royalty basis, to more than two hundred factory installers, including practically every manufacturer of automobile vehicles and the principal manufacturers of industrial equipment. Hydraulic also granted limited manufacturing licenses to General Motors, Ford Motor, and Chrysler Corporation. In 1946, its royalties exceeded $600,000.

Wagner Electric, a general licensee of Hydraulic Brake, supplied practically all of the hydraulic actuating equipment, and Bendix most of the brake assemblies. American automobile manufacturers, except the big three, purchased all their hydraulic braking systems from the two concerns, and even General Motors and Ford Motor bought from them a substantial portion of such systems for their own vehicles.

In 1935, Du Pont entered into an agreement with Wagner Electric to supply Wagner with all its brake fluid requirements. An agreement was also made whereby all Hydraulic Brake licensees except General Motors and Chrysler Corporation had to purchase such fluid either from Wagner or Du Pont.

Bendix organized Bendix-Westinghouse to control also automatic air-braking systems. Bendix received 51 per cent of the stock and

Westinghouse 49 per cent. Each of them gave Bendix-Westinghouse an exclusive license under all of its present and future patents relating to air-braking systems for automotive vehicles. Bendix granted Westinghouse an option to acquire any of its present or future patents pertaining to compressed-air apparatus useful for railway braking systems. As a result of these agreements Westinghouse stayed out of automotive braking and Bendix out of railway. After its organization, Bendix-Westinghouse made direct sales of air-braking systems to manufacturers for their installation and for their service sales. It also made service sales directly to operators of large fleets of motor trucks and to exclusive distributors. Bendix-Westinghouse and Wagner Electric made and sold most of the air-braking systems on automotive vehicles, and they had only one important competitor.

The control of patents, exclusive agreements, and other restrictive measures enabled Bendix, Wagner Electric, and Bendix-Westinghouse to make and sell approximately 80 per cent of the power-braking systems installed on automotive vehicles. In 1946, they manufactured such systems with a total value in excess of $75,000,000.

A district court in a final judgment against Westinghouse in 1948, restricted its control over Bendix-Westinghouse, enjoined its action for infringement of any braking system prior to this date, and ordered royalty-free licenses under seventy-seven patents and patent applications and licenses at reasonable royalties under all its other braking system patents outstanding and issued or applied for within five years.[61]

f. *Glass-Fiber Products.* In the early thirties Owens-Illinois, Corning, and Gustin-Bacon Manufacturing Company each began to develop the commercial manufacture of glass wool, textiles, and mats. In 1935, Owens-Illinois acquired Gustin-Bacon, and in 1938 with Corning, organized the Owens-Corning Fiberglass Corporation of Toledo, Ohio. From 1938 to 1947, this corporation accounted for 98 per cent of the manufacture and 90 per cent of the sales of glass fiber products. According to a government complaint, its alleged monopoly, like that of companies in other parts of the glass industry, rested largely upon patents. Owens-Illinois, in acquiring Gustin-

[61] *United States* v. *Bendix Aviation Corp., Westinghouse Air Brake Co., et al.,* CCH 1948–49 Trade Cases ¶ 62,349 (S.D.N.Y. Civil No. 44–284; Complaint, 1947; Consent Judgment *re* Westinghouse, 1948).

Bacon, had obtained important patents and patent applications. Between 1935 and 1938, Owens and Corning had pooled their patents by cross-licensing each other, and during this period they had obtained patents on the inventions of their employees and also acquired patents and patent rights from outsiders. In 1938 and later, these two companies transferred to Owens-Corning their patents relating to glass-fiber products, and Owens-Corning in turn let the two companies use its inventions outside this field. In government allegations it was pointed out that, in marketing its products, Owens-Corning required purchasers to agree not to manufacture or sell competitive goods and to buy their entire requirements from it.

A consent decree of 1949 enjoined Owens-Corning from acquiring glass-fiber properties from others and ordered Owens-Corning to grant to any applicant royalty-free licenses under 204 patents, with immunity from infringement of them; to grant licenses with reasonable royalties under 60 other patents and any additional patents acquired or applied for within five years; and to reveal "all methods, processes, materials, and machinery" for the commercial production of the inventions in question.[62]

g. *Hosiery Machinery*. The Textile Machine Works and Berkshire Knitting Mills, according to a government complaint, acquired the hosiery machinery of competitors and engaged in other questionable practices. A consent judgment of 1950 ordered the licensing of present and future patents on full-fashioned hosiery royalty-free and on hosiery machinery at reasonable royalties.[63]

h. *Dry Ice and Other Carbon-Dioxide Products*. Solidified into dry ice, carbon dioxide is employed to preserve goods in transportation and storage and to shrink metal parts in airplane and bridge construction. In 1948, the three leading producers and distributors of dry ice were Liquid Carbonic Corporation of Chicago, Wyandotte Chemicals of Michigan, and a subsidiary of Air Reduction Company of New York. Allegations in a government complaint point to their consolidation of dry-ice patents. In 1934, the three companies organized International Carbonic to hold their patents and those of a company called Engineering, but, in 1939, the three companies sold their stock in-

[62] *United States* v. *Owens-Corning Fiberglass Corp.*, CCH 1948–49 Trade Cases ¶ 62,442 (N.D. Ohio Civil No. 5778; Complaint, 1947; Consent Judgment, 1949).

[63] *United States* v. *Textile Machine Works,* CCH 1950–51 Trade Cases ¶ 62,709 (S.D.N.Y. Civil No. 43–671; Complaint, 1947; Final Judgment, 1950).

terest and therefore their patent rights in International Carbonic to Engineering. The sale price of this stock was less than one-seventeenth of its original cost. The apparent purpose of the three companies was to become mere licensees and thus avoid the appearance of unlawful activities. The three companies remained as licensees and paid the same royalties as previously; Engineering continued the monopoly of patents in the dry-ice industry and prevented others from making dry ice by threatening and conducting patent litigation. Licensees were required to reveal their improvements upon the licensed inventions and to transfer relevant patent rights to Engineering. One of the largest manufacturers of ice-making machinery agreed to extend the benefits of its patents to Engineering and to refuse to sell its machinery to others. In addition, the three producers of dry ice allocated markets, conducted price wars against intruders, and purchased all the carbon-dioxide output of several companies.

A consent judgment of 1952 enjoined these purchases and other objectionable practices, forbade International Carbonic from enforcing its principal patent, and required the defendants to license their other patents to applicants at reasonable royalties.[64]

i. *Gas Refrigeration Equipment*. According to a government complaint, Servel, Inc., paid $275,000 for an exclusive license under certain patents relating to gas refrigeration. Through this and kindred agreements Servel acquired approximately 220 patents, and in addition owned about 325 patents obtained from employees. A consent judgment of 1954 required Servel to license, on a reasonable royalty and nonexclusive basis, the approximately 175 patents it then owned and also any others acquired within five years; to furnish at cost technical assistance and information to licensees; and to eliminate restrictive agreements with a Swedish company and foreign manufacturers.[65]

Complaints. Several current complaints point out certain patent consolidations in existence at the present time. Wallace and Tiernan Company monopolized gas-chlorinating equipment by acquiring patents from inventors and others, in addition to acquiring patents from

[64] *United States* v. *The Liquid Carbonic Corp.*, CCH Trade Cases ¶ 67,248 (E.D. N.Y. Civil No. 9179; Complaint, 1948; Final Judgment, 1952).

[65] *United States* v. *Servel, Inc.*, CCH 1953–54 Trade Cases ¶ 67,665 (E.D. Pa. Civil No. 11036; Complaint, 1950; Final Judgment, 1954).

its employees.[66] Magnaflux Corporation sought to acquire through purchase and other means all principal patents relating to magnetic-particle inspection equipment and relating to the powder and paste used in the detection of defects in metal parts.[67] Standard Oil of New Jersey and General Motors assigned to Ethyl Gasoline Corporation their patents and patent applications relating to the use of tetraethyl lead as an antiknock and to the methods of producing the substances.[68] The principal manufacturer of induction-hardened crankshafts purchased patent rights from others and required licensees to grant back any improvements or new developments.[69] International Business Machines systematically acquired developments, inventions, and patents made or owned by others if such developments and inventions might form the basis for a competing line of tabulating machines; it is the sole manufacturer and owner of electrical tabulating machines.[70] Two companies—Westinghouse Air Brake and General Railway Signal—licensed Western Railroad Supply to manufacture and sell railroad and highway grade-crossing devices under patents; they jointly shared the costs of these patents, possessed common rights in them, and received royalties on them.[71]

Is the Control of Practically All Pertinent Patents in an Industry Contrary to the Sherman Act?

Patents for the dominance of an industry may be obtained, to repeat, from independent and hired inventors, manual laborers, and licensees, and the acquisitions may be made, by agreement, even one or more years after the termination of the employment or the license. By these various means, a company may obtain control of improvements upon basic inventions and in turn improvements upon improvements. By controlling the only market for improvement patents and by con-

66 *United States* v. *Wallace and Tiernan Co.*, CCH 1948–49 Trade Cases ¶ 62,305 and ¶ 62,414 (D.C.R.I. Civil No. 705; Complaint, 1946).

67 *United States* v. *Magnaflux Corp.* (S.D.N.Y. Civil No. 51859; Complaint, 1946).

68 *United States* v. *E. I. Du Pont de Nemours and Co., et al.*, CCH 1950–51 Trade Cases ¶ 62,557 and ¶ 62,633 (N.D. Ill. Civil No. 49C–1071; Complaint, 1949).

69 *United States* v. *Ohio Crankshaft Co.* (N.D. Ohio Civil No. 28299; Complaint, 1951).

70 *United States* v. *International Business Machines Corp.* (S.D.N.Y. Civil No. 72–344; Complaint, 1952).

71 *United States* v. *General Railway Signal Co.*, (W.D.N.Y. Civil No. 5237; Complaint, 1952).

trolling the factory operations and laboratories where new and pertinent ideas are most likely to occur, a company can command the stream of inventive thought.

Are these arrangements legal? The judicial interpretation of them is apparently in the making. In 1941, one lower federal court indicated its disapproval. "The allegation is to the effect that the plaintiff bought up practically all patents in the lubricating-equipment business, not for the purpose of marketing the devices covered by the patents, but for the purpose of intimidating and overwhelming and coercing the trade into dealing only with plaintiff. If so, that would constitute a violation of the Sherman Act."[72] In 1942, another lower court, in the Hartford-Empire case, declared, "There should be no doubt but that an aggregation of patents collected into the hands of one or two large concerns, acting in combination with one another for the purpose of attaining control over an important industry in this country, is a violation of the antitrust laws, especially when the means whereby these numerous patents have been acquired are unlawful."[73]

A five to four decision of the Supreme Court, in 1947, in the "Trans-Wrap" case, concerned the legality of an agreement by the licensee to assign improvement patents to the licensor."[74] The issue reached the court after the licensor learned that the licensee had taken out certain patents on improvements in its machine. According to the Supreme Court, "Congress has made all patents assignable and has granted the assignee the same exclusive rights as the patentee," and "so far as the terms of the statute are concerned, we see no difference whether the consideration is services or cash, or the right to use another patent."[75] The licensor, in effect, used one legalized monopoly—a basic patent—to acquire another legalized monopoly—an improvement patent. The end result is the same whether the owner of a basic patent obtains improvement patents by license or purchase.[76]

After emphasizing the legal right to assign a patent, the court indicated possibilities of abuse. "We are quite aware of the possibilities

[72] *Stewart-Warner Corp.* v. *Staley,* 42 F. Supp. 140, 146 (W.D. Pa. 1941).

[73] *United States* v. *Hartford-Empire Co.,* 46 F. Supp. 541, 618 (N.D. Ohio 1942).

[74] *Transparent-Wrap Machine Corp.* v. *Stokes and Smith Co.,* 320 U.S. 637 (1947).

[75] *Ibid.,* 642.

[76] *Ibid.,* 643–45.

of abuse in the practice of licensing a patent on condition that the licensee assign all improvement patents to the licensor. Conceivably the device could be employed with the purpose or effect of violating the antitrust laws. He who acquires two patents acquires a double monopoly. As patents are added to patents a whole industry may be regimented. The owner of a basic patent might thus perpetuate his control over an industry long after the basic patent expired. Competitors might be eliminated and an industrial monopoly perfected and maintained. Through the use of patent pools or multiple licensing agreements the fruits of invention of an entire industry might be systematically funneled into the hands of the original patentee."[77] The court explained that although Congress has not limited the right of the licensee to assign his improvement patents, "that does not mean that the practice we have here has immunity under the antitrust laws. Indeed the recent case of *Hartford-Empire Co.* v. *United States,* dramatically illustrates how the use of a condition or covenant in a patent license that the licensee will assign improvement patents may give rise to violations of the antitrust laws."[78] In the decision the court attempted to rule against abuse but held "that the inclusion in the license of the condition requiring the licensee to assign improvement patents is not *per se* illegal and unenforceable."[79]

One of the dissenting judges said that "the Court's decision in this case unduly enlarges the scope of patent monopolies and is inconsistent with the philosophy enunciated in *Mercoid Corporation* v. *Mid-Continent Co.* and similar cases."[80] Justice Douglas delivered the majority decision. The lower court, to which the "Trans-Wrap" case was remanded for antitrust finding, upheld the assignment provision.[81] However, the lower court found that "the contracting parties were two relatively small concerns and were without an extensive control of market conditions."[82]

Most subsequent cases in which there has been occasion to comment on the "Trans-Wrap" case have emphasized this warning regarding antitrust violation and have found the "grant-back" of im-

[77] *Ibid.,* 646–47.

[78] *Ibid.,* 647–48.

[79] *Ibid.,* 648.

[80] *Ibid.,* 648.

[81] *Stokes and Smith Co.* v. *Transparent-Wrap Machine Co.,* 161 F. 2d 565 (2nd Cir. 1947).

[82] *United States* v. *General Electric Co.,* 80 F. Supp. 989, 1005 (S.D.N.Y. 1948).

provements in their particular cases to be illegal. In 1948, the Federal Trade Commission ordered Pure Carbonic and its associates to cease and desist from owning or controlling the patent rights or patent applications relating to the manufacture of solid carbon dioxide by which competing or prospective manufacturers are hindered from continuing or beginning the production of this product.[83] In the incandescent lamp case of 1949 a district court declared, "In this present case, General Electric regimented an industry by, among other things, its acquisition of patents to perpetuate a control over the incandescent electric lamp, long after its basic patents expired, to maintain a dominant position rendering it possible for it to eliminate competition and maintain an industrial monopoly of the type recognized by the Transparent-Wrap case to be an eventuality violative of the antitrust laws. Its aggregation of patents into its control permitted General Electric to monopolize patents and by so doing it violated Section 2 of the Sherman Antitrust Act."[84]

As a result of Supreme Court decisions in three cases, holding in effect that Section 7 of the Clayton Act did not prohibit asset acquisitions, the Federal Trade Commission in its annual report for the year 1927 first recommended the enactment of legislation to close this loophole in the law. Twenty-three years later, in 1950, an amendment to this section curbed the legal right to consolidate assets. It provides that "no corporation subject to the jurisdiction of the Federal Trade Commission shall acquire the whole or any part of the assets of another corporation engaged also in commerce, where in any line of commerce in any section of the country, the effect of such acquisition may be substantially to lessen competition or to tend to create a monopoly." Will this amendment be applied to the acquisition of patents as it has been to physical assets? Logic and experience justify an affirmative answer.

Conclusion

Until recently at least, one company may acquire the patents of another or a third may obtain the patents of both without any violation of the law. However, restraint of trade between even two of

[83] *Federal Trade Commission* v. *Pure Carbonic, Inc.*, F.T.C. Annual *Report,* 1948; Complaint, 5143 p. 44.

[84] *United States* v. *General Electric Co.*, 82 F. Supp. 753, 816 (D.N.J. 1949).

them has been condemned many times by the Supreme Court, especially within recent years. The same contrast applies to physical assets as well as to patents and other intangible property. However, the amendment of the Clayton Act just noted, and the recent consideration by the courts of antitrust laws in interpreting the consolidation of competing patents, may mark a definite reversal of the trend toward the ownership of important industries by one or a few companies.

In 1945, the Supreme Court, in reviewing a lower court's antitrust decree for Hartford-Empire, approved compulsory licensing with royalties but opposed it if without royalties. In spite of this opposition, several companies, beginning in 1946, accepted consent decrees which required them either to dedicate certain patents to the public or to grant licenses to any applicant without compensation.

CHAPTER IV

License Agreements

License agreements may be classified in several ways—particularly according to subject matter, activity of members, number of participants, and contractual consideration. In a general way they are based on the right to exclude others from making, using, and selling. As already indicated, the patent consolidation, also the patent pool, may extend and strengthen its control by license agreements, particularly those agreements which require "grant-backs" of improvements and which prescribe the kind or quantity of production. Some license agreements stipulate the use to which the invention may be put or the industry in which it may be employed or the supplies with which it may be used; the attempts of patent owners to control supplementary products are extensive enough to justify an entire chapter devoted to a discussion of them. However, the most significant phase of license agreements pertains to the sale of the invention—usually the price or market—and the present chapter is devoted to that.

The licensor may be engaged in the manufacture and sale of the product which its patents cover, or, like Hartford-Empire, it may be primarily or only in the business of amassing patents and then licensing others. The licensees may include only manufacturers, or they may include both manufacturers and middlemen and therefore permit both horizontal and vertical restraints of trade. The license agreement may also be single or multiple in that it may involve one or more patent owners and one or more licensees; two or more patent owners, acting in unison in dealing with two or more licensees, usually engage in the most significant and questionable relationships. The licensees, in return for the privileges they receive, may pay royalties, accept the validity of the patents, grant back their improvements, and perhaps comply with other terms. Of course, each license agreement is concerned to some extent with all the phases of licensing—subject matter, activity, number and contract consideration.

Many interferences in the Patent Office and suits for infringement and issues of validity are settled by granting licenses. A license usually provides that if more favorable terms are given to anyone else they shall apply also to the other licensees. The agreement may not enumerate particular patents, but instead cover rights in certain subject matter. This is typical of licenses granted or taken by large corporations in their licensor-licensee relationships. Even if the patents are enumerated, licensees do not know which ones they are using, so vague and overlapping they may be. Licensees pay royalties for the right of making and using and selling the invention and therefore of avoiding possible infringement, and in some instances they pay for the privilege of restraining trade presumably without hindrance. In many instances both the licensor and licensee are former competitors.

This patent controlling and licensing system exists in numerous industries and apparently is the most effective scheme for monopolizing and restraining trade. It has been brought before the courts by the government and in private litigation. From the standpoint of prosecution by the Department of Justice and interpretation by the courts, the history of license agreements relating to the sale of a patented product, like the history of consolidations, may be divided into two parts—indifference and leniency to about 1940 and vigorous prosecution and stricter interpretation after that time. After presenting cases which illustrate these two periods, license agreements with middlemen only will be presented.

Indifference and Leniency, to 1940

From 1902 to about 1940 the courts seemed to interpret license agreements primarily from the standpoint of the historic right of the patentee to grant a license with restrictions on the making, using, and selling of his inventions. As already noted, the Supreme Court in 1902 upheld the agreement between National Harrow and Bement on the basis of this right to grant a license to another upon certain conditions. And the Supreme Court approved the licenses in the gasoline cracking case on the ground that the licenses were open to all and involved a low percentage of interstate commerce in gasoline. Other lenient interpretations, which do not involve patent pools, pertain to rubber tires, wind stackers, incandescent lamps, and amplifiers.

Rubber Tires. The Rubber Tire Wheel Company, the owner of a patent on rubber tire wheels, entered into uniform license agreements with eighteen—practically all—of the manufacturers of tires in the United States. The contract provided, among other things, for a considerable advance in the selling price of tires and limitation of each licensee's production to a certain percentage of the output of all. The patent owner, soon after the formation of this pool, brought suit against one of the licensees to recover unpaid royalties. In 1906, a circuit court declared the license agreement illegal owing to the pronouncement by the Sixth Circuit Court of Appeals of the invalidity of the patent in question and owing to the creation of a fund to promote the scheme of underselling and, hence, crushing competitors.[1] According to law, this decision defining the invalidity was effective only in the sixth circuit.

In 1907, the case, appealed from the circuit court to the Seventh Circuit Court of Appeals, received a very different interpretation. After elaborating the idea that a patentee enjoys an exclusive monopoly, the higher court asked these questions: "Over and above an absolute monopoly created by law, how can there be a further and an unlawful monopoly in the same thing? If plaintiff were the sole maker of Grant tires, how could plaintiff's control of prices and output injure the people, deprive them of something to which they have a right? Is a greater injury or deprivation inflicted, if plaintiff authorizes a combination or pool to do what plaintiff can do directly?"[2]

But the plaintiff alone could not directly secure a monopoly of rubber tires. Its patent was only one of many outstanding and pertaining to rubber tires. Its insignificant and questionable nature is suggested by the announcement of its invalidity by another court. The argument of the Seventh Circuit Court of Appeals would be sound if the plaintiff had held a patent monopoly of the rubber tire as a commodity—a proviso contrary to fact, for there were at least eighteen other manufacturers of tires. The scheme brought rubber tires as such, as well as the particular invention relating to rubber tires, within the realm of a monopoly. The acceptance of the appellate court's argu-

[1] *Rubber Tire Wheel Co.* v. *Milwaukee Rubber Works Co.,* 142 Fed. 531, 532 (C.C.E.D. Wis. 1906).

[2] *Ibid.,* 154 Fed. 358, 363 (7th Cir. 1907).

ment would lead to a monopoly of every product for which one or more patents had been issued. It would only be necessary for one manufacturer who had a patent relating to a certain product to license other manufacturers of the same product upon certain conditions with respect to price, as the Rubber Tire Wheel Company did. For example, a patent covering a pocket knife, no matter how trivial, might be made the legal basis of a combination of all manufacturers of pocket knives. The failure to distinguish between a product and a patent relating to the product accounted for the weakness of the court's argument. As illustrated later, such defective reasoning is found in other decisions. This case reached the Supreme Court, where it was dismissed, per stipulation.[3]

Wind Stacker. The Indiana Manufacturing Company, which gradually acquired practically all patents pertaining to wind or pneumatic straw stackers, issued uniform licenses to all the manufacturers of threshing machines in the country.[4] Manufacturers were given the right to use any and all inventions covered by such patents and by any other patents subsequently acquired. The company stipulated a uniform price for its product and the payment of a royalty. It brought suit against Case Manufacturing Company for infringement of its patents and violation of the license agreement between the two companies.

A circuit court, in 1906, pronounced Indiana Manufacturing the promoter of a two-fold combination of numerous patent properties and of all manufacturers of threshing machines, which made its agreement with the licensees illegal. This court considered, therefore, not only the relation existing between the plaintiff and the single defendant, as in the Bement case, but also the general relation of the complainant or licensor to all its licensees. It announced in a forceful way that the patent statute does not furnish the foundation for a combination of the manufacturers of any single commodity. It declared that it would not consider the infringement suit in question since it grew out of an illegal agreement.[5]

A circuit court of appeals, in 1907, rendered a decision diametrically opposed to that of the circuit court. The judge of the appellate

[3] 210 U.S. 439 (1908).

[4] *Farm Machinery Trade Associations*, 122–23.

[5] *Indiana Mfg. Co.* v. *J. I. Case Threshing Machine Co.*, 148 Fed. 21, 25–26 (C.C.E.D. Wis. 1906).

court who wrote the opinion was the same one as in the rubber tire case, and his argument was much the same; both decisions were promulgated the same day. The court stated that articles covered by patents are never articles of trade or commerce within the operation of the Sherman law. It attempted to answer the objection to the large number of licensees: "All the makers of threshing machinery have come into the system. That this resulted, without any concert of action on the part of the licensees, solely from a policy pursued by appellant through a course of years, is virtually admitted and is clearly proven. Appellant started out to supply the trade. That was its exclusive right."[6] The court declared that the number of patents owned by Indiana Manufacturing Company was irrelevant.

The case, after an appeal to the Supreme Court, was "dismissed without costs to either party, per stipulation."[7] In the interim—between the appeal and the dismissal—the directors of Case had purchased a majority of the stock of Indiana Manufacturing, a fact which partly explains the interest of both parties in dropping the suit.

Shoe Lasts. The Krentler-Arnold Hinge Company, the owner of patented shoe and boot lasts, required its licensees to maintain the prices of all lasts, both patented and unpatented, in accordance with a schedule which it provided. A district court, in 1913, enjoined them from "maintaining or observing an agreed price upon unpatented lasts, parts, or fittings," but thus, in effect, permitted this price policy for such articles if patented.[8]

Incandescent Lamps. General Electric granted a license to Westinghouse which required the latter to adopt the prices and plan of distribution which GE prescribed. In interpreting this relationship in 1926, the Supreme Court pointed out that GE "is the owner of three patents—one of 1912 to Just and Hanaman, the basic patent for the use of tungsten filaments in the manufacture of electric lamps; the Coolidge patent of 1913, covering a process of manufacturing tungsten filaments by which their tensile strength and endurance are greatly increased; and, third, the Langmuir patent of 1916, which is for the use of gas in the bulb by which the intensity of the light is substantially heightened. These three patents cover completely

6 *Ibid.*, 154 Fed. 365, 370 (7th Cir. 1907).

7 207 U.S. 603 (1907).

8 *United States* v. *Krentler-Arnold Hinge Last Co.*, 1 D & J. 407, 409 (1913).

the making of the modern electric lights with the tungsten filaments, and secure to General Electric the monopoly of their making, using, and vending."[9] In reply to its own question as to whether a patentee may grant licenses that "limit the selling by limiting the method of sales and price," the Supreme Court stated, "We think he may do so, provided the conditions of sale are normally and reasonably adapted to secure pecuniary reward for the patentee's monopoly."[10] In appraising this stand in the light of its denial of the right of a patentee to prescribe supplies for his invention the court said, "The price at which a patented article sells is certainly a circumstance having a more direct relation, and is more germane to the rights of the patentee, than the unpatented material with which the patented article may be used. Indeed, as already said, price fixing is usually the essence of that which secures proper reward to the patentee."[11] A second issue in the case, involving the legality of so-called agency contracts with distributors, is presented later in this chapter as a phase of resale price maintenance.

Amplifiers. In 1938, the Supreme Court decided that a licensee could be legally restricted only to one field in using or applying an invention. In this instance amplifiers embodying the invention were employed both in talking-picture equipment and also in radio-broadcast reception. A patent pool granted an exclusive license for talking pictures to Western Electric and nonexclusive licenses for radios to about fifty manufacturers of radio equipment. One of the fifty was American Transformer Company. This company, knowing that it had not been licensed to manufacture or to sell amplifiers for use in talking-picture equipment of theaters, made and sold them to Pictures Corporation, which knew of this restriction upon American Transformer. The Supreme Court held that both of them were infringers.[12]

This decision, six to two, cites and quotes with approval the General Electric case of 1926. Justices Black and Reed, who dissented, were two of the four judges that later neutralized this case in the Line Material decision of 1948.[13] Justice Roberts took no part in the decision. The court added, "We have consequently no occasion to con-

[9] *United States* v. *General Electric Co.,* 272 U.S. 476, 480–81 (1926).
[10] *Ibid.,* 490.
[11] *Ibid.,* 493.
[12] *General Talking Pictures Corp.* v. *Western Electric Co.,* 305 U.S. 124 (1938).
[13] *United States* v. *Line Material Co.,* 333 U.S. 287 (1948).

sider what the rights of the parties would have been if the amplifier had been manufactured 'under the patent' and 'had passed into the hands of a purchaser in the ordinary channels of trade.' Nor have we occasion to consider the effect of a 'licensee's notice' which purports to restrict the use of articles lawfully sold."[14] Apparently these remarks were designed to avoid any appearance of conflict with other decisions of this tribunal which limit control over licensees and buyers as to prices and supplies. This is one of several instances in which the Supreme Court has difficulty in reconciling the controls which a patentee may and may not exercise over his licensees.

Vigorous Prosecution and Stricter Interpretation after 1940

Beginning about 1940 there was a significant increase in the antitrust activity of the Department of Justice, and the courts, in interpreting the relation of license agreements for the sale of products to antitrust legislation, emphasized the combined effect of license agreements upon competition rather than emphasizing the right to grant a single license with restrictions. In fact, the forties and fifties constitute a turning point in the appraisal of various uses of patents and in the approval of compulsory licenses and even of the dedication of patents to the public.

Leaded Gasoline. Ethyl Gasoline Corporation, one-half owned by Standard Oil of New Jersey and the other half by General Motors and Du Pont, had two patents on a poisonous fluid containing lead, a third patent on the mixtures, and a fourth one on a method of using it. This Ethyl company manufactured the fluid and sold it outright, without royalties, to refiners and jobbers. It derived its profit solely from the sale of the patented fluid.[15]

Refiners mixed this fluid with their gasoline and sold the resultant patented mixture through their own wholesale and retail outlets, to individual retailers and consumers directly, and to jobbers who received it at their bulk storage plants and then sold it to retailers for distribution to consumers. The refiners were required by license agreements to sell only to jobbers licensed by Ethyl; to maintain a prescribed price differential between regular and ethyl gasoline; to

[14] 305 U.S. 124, 127 (1938).
[15] *Ethyl Gasoline Corp.* v. *United States,* 309 U.S. 436, 446 (1940).

comply with certain conditions in the use of the name Ethyl; and to conform to United States health regulations in mixing the fuel.[16]

Jobbers desiring to sell ethyl gasoline had to secure licenses from the Ethyl company through the refiner from which they expected to purchase motor fuel. They had to sell and deliver within a specified territory regular and ethyl gasoline from a designated licensed refiner; to furnish monthly a list of all places at which the motor fuel was sold; to comply with health regulations relating to the handling of the motor fuel; to refrain from adulterating or diluting the fuels; and to use the Ethyl name only under certain conditions. Their licenses could be cancelled for failure to comply with these restrictions.[17] Some of them believed that the licensing practices of Ethyl required them "to maintain prices and abide by the marketing practices of the major oil companies."[18]

The Supreme Court declared that "by the authorized sales of the fuel by refiners to jobbers the patent monopoly over it is exhausted," and after the sale neither Ethyl nor the refiners "may longer rely on the patents to exercise any control over the price at which the fuel may be re-sold."[19] The Supreme Court believed that "agreements for price maintenance of articles moving in interstate commerce are . . . unreasonable restraints within the meaning of the Sherman Act because they eliminate competition."[20] The court thus took a stand against the extension of the patent monopoly to avoid competition among buyers.

Apparently the Ethyl company relied greatly upon the appearance of protecting the public health in order to justify the control over the prices and other marketing practices of the jobbers. The tribunal pointed out, "There is no authentic instance of injury resulting from the handling of lead-treated gasoline, after its manufacture, attributable to its lead content."[21] In any event, the court concluded that any necessary protection of the public from leaded gasoline and adulterated fuel might be given "without resort to the jobber license device which has been and is capable of being used for other and illicit purposes."[22]

16 *Ibid.*, 447–48.
17 *Ibid.*, 448–49.
18 *Ibid.*, 454.
19 *Ibid.*, 457.
20 *Ibid.*, 458.
21 *Ibid.*, 460.

Hardboard. A recent and interesting instance of price fixing through licenses to vend or through so-called agencies involved ten companies —Masonite, Celotex, Certain-Teed Products, Johns-Manville Sales, Insulite, Flintkote, National Gypsum, Wood Conversion, Armstrong Cork, and Dant and Russell. Each of them engaged either in manufacturing and selling building materials or in selling building materials manufactured by others. All maintained selling organizations and to a great extent competed in the same markets. Some of them had competing patents, while others did not.[23]

The price-fixing scheme centered around Masonite. It manufactured, from wood chips, a synthetic hardboard used in the building industry and distributed the product through its own selling organization. Between 1926 and 1928, four patents were issued to Masonite, the claims of which covered both hardboard and the processes for making it. In 1929, Celotex, to which several patents were issued, began to make hardboard out of the waste material which results from the grinding of sugar cane. In 1931, Masonite, after refusing to enter into a cross-licensing agreement with Celotex, brought suit against this company for infringement of one of its patents. A circuit court of appeals held that Masonite's patent was both valid and infringed.[24] Negotiations between these companies led to the settlement of the patent litigation and the execution of the so-called "agency" agreement of 1933. Masonite executed an identical agreement with six other companies by June, 1934, and in each instance it informed the other party of a similar agreement with the other companies. It brought an infringement suit against a dealer who handled Insulite's hardboard. Negotiations before trial resulted in an agency agreement substantially similar to the others. Disputes between Masonite and these "agents" led, in 1936, to a modification of their agreements. The new contract was signed by each of the companies and became effective only when all the so-called agents had agreed to it. In 1937, two more companies became agents.[25]

According to the agreements, Masonite designated each of the other nine companies as an agent. It fixed, from time to time, the minimum selling price of its product and the maximum terms and

[22] *Ibid.*, 459.
[23] *United States* v. *Masonite Corp.*, 316 U.S. 265, 267 (1942).
[24] *Ibid.*, 268.
[25] *Ibid.*, 269–70.

conditions of sale for its agents and itself. The agreement provided that on direct shipments to agents the hardboard "shall be received and held on consignment" and "title thereto shall remain" in Masonite until sold by the "agent."[26] The minimum prices were f.o.b. Masonite's factory. The agent paid not only freight but also taxes and insurance, received specified commissions on each sale and sold only to the construction industry, the industrial market being reserved for Masonite. The agent compensated Masonite "by advancing one-half of the difference between the list price and the agent's discount within twenty days after the close of the month in which the order was shipped, and the balance within twenty days after the close of the month in which the products were sold."[27] This provision especially, it should be mentioned here, indicates a hybrid agent—one who pays partly before and partly after the sale of the goods.

In 1942, the Supreme Court held that the pattern of relationship between Masonite and the other companies "was fixed in 1933 and its fundamental characteristics were maintained, not basically altered, in 1936."[28] It quoted its own words in the Interstate Circuit case: "Acceptance by competitors, without previous agreement, of an invitation to participate in a plan, the necessary consequences of which, if carried out, is restraint of interstate commerce, is sufficient to establish an unlawful conspiracy under the Sherman Act."[29] The court felt that "the fixing of prices by one member of a group, pursuant to express delegation, acquiescence, or understanding, is just as illegal as the fixing of prices by direct, joint action";[30] and it cannot be justified because "the effect of the combination may have been to increase the distribution of hardboard, without increase of price to the consumer, or even to promote competition between dealers."[31]

Patents did not save the price-fixing scheme. The court said, "The owner of a patent cannot extend his statutory grant by contract or agreement."[32] It rejected the proposition that "since the patentee has the power to refuse a license, he has the lesser power to license on

26 *Ibid.*, 271.
27 *Ibid.*, 271–72.
28 *Ibid.*, 274.
29 *Ibid.*, 275.
30 *Ibid.*, 276.
31 *Ibid.*, 276.
32 *Ibid.*, 277.

his own conditions."[33] The court held in this instance that the patented product had passed into the hands of the purchaser and therefore was no longer under the protection of the patent law. It swept away the pretense of an agency relationship. "So far as the Sherman Act is concerned, the result must turn not on the skill with which counsel has manipulated the concepts of 'sale' and 'agency' but on the significance of the business practices in terms of restraint of trade. . . . In this case, the price regulation was based on mutual agreement among distributors of competing products, some of whom had competing patents."[34] This is a step toward the later doctrine in the Line Material decision of limiting rights if there are several licensees.

Furnace Control. The Minneapolis-Honeywell Regulator Company owned a combination patent on a system of furnace control which required three thermostats for its operation. It licensed five of its manufacturing competitors under this patent. The licenses established minimum prices for the sale of the controls, and those prices could not be cut by the licensees through the inclusion of "extras" or through the reduction of charges for services. Price lists were attached governing sales to manufacturers, jobbers, wholesalers, and dealers. Equal terms to all licensees were provided. One of the licensees, charged with infringement, contended that the price restriction in the licenses violated the Sherman Act. However, the judicial interpretation of it turned upon the issue of control of an unpatented part of the combination invention rather than license restrictions: "The effort here made to control competition in this unpatented device plainly violates the antitrust laws, even apart from the price-fixing provisions of the license agreements."[35]

Parking Meters. According to testimony offered in a district court in 1944, a promoter wrote, "In order that a price structure may be set up in this country in parking meters, it is absolutely essential that all important manufacturers be brought under a single organization set up to control prices. That is the function of Vehicular Parking."[36] Six manufacturers assigned any patents they had to this

[33] *Ibid.*, 277.
[34] *Ibid.*, 280.
[35] *Mercoid Corp.* v. *Minneapolis-Honeywell Regulator Co.*, 320 U.S. 680, 684 (1944). For further discussion, see Chapter Six.
[36] *United States* v. *Vehicular Parking*, 54 F. Supp. 828, 839 (D. Del. 1944).

company and accepted license agreements under restrictive conditions. Infringement suits against three of them and interference in the Patent Office with another helped to bring all of them to terms. It was agreed that the parking-meter industry would be limited to this group.[37] An outsider wrote in 1940: "I would, therefore, like to know if a license to manufacture under your group of patents would be available to me, and what the cost of the same would be, and under what patents you can license." Vehicular replied, "There is no plan to grant additional licenses at the present time."[38] The license agreements, policed by Vehicular, covered prices, terms, and conditions for the sale of parking meters, parts, services, and accessories. The licensees controlled from 95 to 98 per cent of the parking-meter business.[39]

The courts held that the government had sustained its charges of violation of the Sherman Act. Decrees of 1944, 1945, and 1946 ordered Vehicular and its licensees to issue to any applicant an absolutely unrestricted license under their pooled patents, but permitted them to charge a reasonable royalty after the entry of the order in 1946. The defendants were further enjoined from bringing suits for past infringement or unpaid royalties, and from disclosing to stockholders information about parking meters emanating from present or prospective purchasers of them unless such information was also made available to all other manufacturers of parking meters.[40]

The Vehicular case reveals a company controlling the patents of an industry through purchase and license, uniform licenses regulating price to former competitors, refusal of licenses to others, control of unpatented parts, intimidation through infringement suits, and patents of questionable validity—all for monopoly and restraint of trade.

Gypsum Board. The United States Gypsum Company, since its organization in 1901, has been the dominant concern in the gypsum industry. It acquired the most important patents relating to the manufacture of gypsum board and granted licenses to other concerns in the industry. In some instances patent litigation led to assignment of

[37] *Ibid.*, 831–33.

[38] *Ibid.*, 832.

[39] *Ibid.*, 833–34.

[40] *Ibid.*, 54 F. Supp. 828 (D. Del. 1944); 56 F. Supp. 297 (D. Del. 1944); and 61 F. Supp. 656 (D. Del. 1945); also, CCH 1944–45 Trade Cases ¶ 57,226, ¶ 57,277 and ¶ 57,404.

patents to it and its grant of licenses in turn to their former owners. National Gypsum, Certain-teed Products, Celotex, Ebsary Gypsum, Newark Plaster—these companies and one individual became licensees. In 1939, United States Gypsum sold 55 per cent of all gypsum board in the area east of the Rocky Mountains; National Gypsum, 23 per cent; Certain-teed, 11 per cent; and the other four companies, smaller percentages. These companies had acquired several other concerns prior to becoming licensees.[41]

License agreements, effective in 1929, were substantially identical in terms. They covered fifty patents and seven patent applications, and each was to run until the most recent patent expired. United States Gypsum reserved the right to fix the minimum price at which each licensee sold wallboard embodying its patents, and it agreed that this price would be no greater than its own price. The licensees agreed not to sell patented wallboard to manufacturing distributors unless this company gave its consent concerning each prospective purchaser. Each licensee agreed to pay as royalty a stipulated percentage of the selling price of all plaster board and gypsum wallboard of every kind, whether or not covered by patents.[42]

United States Gypsum issued a series of bulletins which defined in minute detail both the prices and terms of sale for patented gypsum board. In 1930, it ordered the elimination of the jobbers' discount. It controlled the price of plaster, though unpatented, so that cutting its price could not be used indirectly for cutting the price of gypsum board—these two products are usually sold together. It adopted a basing-point system of pricing like the one in the cement industry; each licensee quoted a price determined by taking the mill price at the nearest basing point and adding the all-rail freight charge from that point to destination. The freight was computed on specified uniform weights in order to prevent variations in freight arising from differences in weight of board made by different manufacturers. Each licensee charged exactly the same switching, cartage, and extra delivery charges. Sales on consignment and long-term credit were forbidden. In other words, standardization was effected to facilitate the control of prices.[43]

In 1940, the government brought suit against United States Gyp-

[41] *United States* v. *United States Gypsum Co.,* 333 U.S. 364, 368–69 (1948).
[42] *Ibid.,* 381.
[43] *Ibid.,* 382–83.

sum and its six licensees for violation of the Sherman Act. In 1948, the Supreme Court declared, "We think that the industry-wide license agreements, entered into with knowledge on the part of licensor and licensees of the adherence of others, with the control over prices and methods of distribution through the agreements and the bulletins, were sufficient to establish a *prima-facie* case of conspiracy."[44] The license agreements themselves, according to the court, revealed the purpose. "The licensor was to fix minimum prices binding both on itself and its licensees; the royalty was to be measured by a percentage of the value of all gypsum products, patented or unpatented; the license could not be transferred without the licensor's consent; the licensee opened its books of accounts to the licensor; the licensee was protected against competition with more favorable licenses and there was cancellation clause for failure to live up to the arrangements."[45]

The defendants maintained that earlier decisions of the Supreme Court had recognized the legal right of a patentee to stipulate the prices at which his licensees may manufacture and sell. However, the Supreme Court disagreed. "The General Electric case affords no cloak for the course of conduct revealed in the voluminous record in this case. That case gives no support for a patentee, acting in concert with all members of an industry, to issue substantially identical licenses to all members of the industry under the terms of which the industry is completely regimented, the production of competitive unpatented products suppressed, a class of distributors squeezed out, and prices on unpatented products stabilized."[46] Still more significant is the next statement. "Even in the absence of the specific abuses in this case, which fall within the traditional prohibitions of the Sherman Act, it would be sufficient to show that the defendants, constituting all former competitors in an entire industry, had acted in concert to restrain commerce in an entire industry under patent licenses in order to organize the industry and stabilize prices. That conclusion follows despite the assumed legality of each separate patent license, for it is familiar doctrine that lawful acts may become unlawful when taken in concert. Such concerted action is an effective deterrent to competition."[47]

[44] *Ibid.*, 389.
[45] *Ibid.*, 389.
[46] *Ibid.*, 400.
[47] *Ibid.*, 401.

Neutral Vote on General Electric Decision of 1926. In passing upon the legality of the price-fixing provision in the Line Material case, the Supreme Court voted four to four on whether to accept the General Electric decision of 1926 as a precedent.[48] Justice Douglas, in speaking for himself and three other justices, said that the Supreme Court in that decision "saddled the economy with a vicious monopoly."[49] According to him, the patentee and his associates "get the benefits of a conspiracy or combination in restraint of competition. That is more than an 'exclusive right' to an invention; it's an 'exclusive right' to form a combination with competitors to fix the prices of the products of invention. The patentee creates by that method a powerful inducement for the abandonment of competition, for the cessation of litigation concerning the validity of patents, for the acceptance of patents no matter how dubious, for the abandonment of research in the development of competing patents. Those who can get stabilized markets, assured margins, and freedom from price cutting will find a price-fixing license an attractive alternative to the more arduous methods of maintaining their competitive positions. Competition tends to become impaired not by reason of the public's preference for the patented article but because of the preference of competitors for price fixing and for the increased profits which that method of doing business promises."[50]

In this same opinion, Justice Douglas said, "To allow these price-fixing schemes is to reverse the order and place the rewards to inventors first and the public second. This is not the only way a patentee can receive a pecuniary reward for his invention. He can charge a royalty which has no relation to price fixing. Or he can manufacture and sell at such price as he may choose. Certainly if we read the patent statutes so as to harmonize them as closely as possible with the policy of antitrust laws, we will strike down a combination which is not necessary to effectuate the purpose of the patent statutes. If we did that in this case we would overrule the General Electric case."[51]

Justice Douglas also pointed out that the Supreme Court, not Congress, was the author of the doctrine followed in the GE case. "Since we created it, we should take the initiative in eliminating it. It is hard

[48] *United States* v. *Line Material Co.*, 333 U.S. 287 (1948).
[49] *Ibid.*, 318.
[50] *Ibid.*, 319.
[51] *Ibid.*, 320–21.

for me to square it with the standards which the Constitution has set for our patent system. It plainly does violence to the competitive standards which Congress has written into the Sherman Act."[52] Justices Black, Murphy, and Rutledge agreed with this opinion. Justices Harold H. Burton, Frank M. Vinson, and Frankfurter still accepted the GE doctrine and construed it as applicable in this case. Justice Reed favored the doctrine but held it inapplicable here because of multiple licensees. Justice Robert H. Jackson did not participate.

Incandescent Lamps. After the expiration of its basic patents, the last one in 1933, GE continued its control of incandescent lamps. In 1941, the Department of Justice filed a complaint against it, Westinghouse, and several other manufacturers for violation of antitrust laws. The trial, commenced in 1942, was postponed in 1943, at the request of the War and Navy departments because GE maintained that it interfered with war work. The trial was resumed in 1946 and a district court reached a decision in 1949.[53]

At the time the suit was filed, Westinghouse had an "A" license, and Sylvania and four smaller companies had "B" licenses all of which were renewals of licenses granted prior to 1922. The "A" license established price and quota restrictions, cross-licensing, and exchange of know-how. The "B" licenses contained no price restrictions but limited production to a percentage of GE sales.[54] The growth of its licensees "was carefully limited to fixed percentages of its own production and expansion so that over the years its share of the business was not materially diminished and its dominant proportion was never exposed to any hazard in that direction."[55] The fixed quotas on sales of "B" licensees restricted their growth and thus "promoted the domination of General Electric over the incandescent lamp industry."[56] From 1927 to 1940, GE's average percentage of the total sale of all lamps was 55 per cent; Westinghouse and its other manufacturing licensees accounted for 27.2 per cent. GE's lamp department enjoyed substantial profits—a rate of 22.8 per cent after taxes even from 1930 to 1939.[57]

[52] *Ibid.*, 321.
[53] *United States* v. *General Electric Co.,* 82 F. Supp. 753 (D.N.J, 1949).
[54] *Ibid.,* 776–77.
[55] *Ibid.,* 905.
[56] *Ibid.,* 858.
[57] *Ibid.,* 893–95.

Westinghouse obtained other concessions that were denied to the "B" licensees. "Under the Corning–General Electric agreements, it received the benefits accruing therefrom. It was the favored beneficiary of all that flowed from employment of the 'Mazda' symbol. It was granted a share in the foreign lamp markets by virtue of the agreements negotiated by International General Electric. Granted a larger share of the domestic lamp business, its policy and program were those of General Electric since both operated under the 'Mazda' symbol."[58] It was "in the status of a controlled competitor."[59] GE, for example, in urging upon cities the "Mazda" mark as a specification for incandescent lamps, was in effect "reducing the market to the only two manufacturers under the 'Mazda' mark and these had the relationship of licensor-licensee."[60]

According to the court, "General Electric's apparently impregnable position was formidable barrier to anyone who contemplated entering the lamp manufacturing field and this, coupled with the knowledge that it controlled the manufacture of lamp bases, lamp manufacturing machinery, along with a tight block on the supply of glass, created a situation sufficient to deter entry. The link to unlawful monopoly is apparent from the fact that upon expiration of the lawful patent monopoly in 1933, there was no new entry into the field and there were no new General Electric licenses."[61] It should be noted here that GE and Corning manufactured all the bulbs for lamps made in the United States. Corning, licensed under GE's patent covering the Pipkin invention on inside frosted bulbs, sold them only to GE licensees. This patent, though upheld by two lower courts, was declared invalid by the Supreme Court in 1945—after its expiration.[62]

A final judgment of a district court, in 1953, provided that GE and other defendants must dedicate to the public their patents on incandescent lamps and lamp parts; that GE must license its lamp machinery patents at reasonable royalties and all its patents issued to it within five years on such lamps, parts, and machinery at reasonable royalties; and that certain objectionable contracts between GE and other companies be cancelled. In support of this stand, Judge

[58] *Ibid.*, 873.
[59] *Ibid.*, 850.
[60] *Ibid.*, 852.
[61] *Ibid.*, 900.
[62] *Ibid.*, 792.

Phillip Forman quoted from the Supreme Court in the National Lead Case, "While it has been contended that because of the decision of this Court in *Hartford-Empire Company* v. *United States*, 323 U.S. 386, the District Court was not free in the present case to require the issuance of royalty-free licenses, we feel that, without reaching the question whether royalty-free licensing or a perpetual injunction against the enforcement of a patent is permissible as a matter of law in any case, the present decree represents an exercise of sound judicial discretion."[63] He then added, "The foregoing language certainly casts a shadow on the ruling in the Hartford-Empire case as being a hard and fast determination against royalty-free licenses, and suggests, if indeed it does not invite, the application of such a measure when and where the circumstances of patent abuse prescribe it."[64] Westinghouse and Corning, in 1942 and 1946 respectively, accepted consent decrees which provided, among other things, for royalty-free licenses.[65]

Wrinkle Finishes for Paint. Two companies assigned all their patent rights relating to wrinkle paint finishes to New Wrinkle Inc. in return for its common and preferred stock. This new company, solely a patent holder, granted in turn nonexclusive licenses to them and other paint manufacturers which prescribed minimum prices for all of them. A district court in Ohio dismissed the case on the ground that New Wrinkle had never been engaged in interstate commerce and therefore could not be guilty of violating the Sherman Act. However, upon appeal, the Supreme Court held that "this making of license contracts for the purpose of regulating distribution and fixing prices of commodities in interstate commerce is subject to the Sherman Act, even though the isolated act of contracting for the licenses is wholly within a single state." In this situation, as in the Line Material and Gypsum cases, "an arrangement was made between patent holders to pool their patents and fix prices on the products for themselves and their licensees."[66]

[63] *United States* v. *National Lead Co.*, 332 U.S. 319, 338 (1947).

[64] *United States* v. *General Electric Co.*, 115 F. Supp. 835, 843 (D.N.J. 1953).

[65] *United States* v. *General Electric Co., Westinghouse Electric Mfg. Co., et al.*, CCH Trade Regulation Reports, Supp. 1941–43 ¶ 52,777 (D.N.J. Civil No. 1364; Complaint, 1941; Consent Decree *re* Westinghouse, 1942). The Corning decree is cited below in footnote 74.

[66] *United States* v. *New Wrinkle, Inc.*, 342 U.S. 371, 377, 380 (1952).

Other Products. Complaints and consent decrees indicate the following examples of license agreements in restraint of trade.

a. *Heating and Air-conditioning Equipment.* Aerofin Corporation, organized and owned by three manufacturing companies, had a patent on encased coils used for heating and air-conditioning equipment. According to a government complaint, Aerofin licensed about twenty-five other manufacturers and distributors, prescribed the prices for these coils, and required the manufacturers to pay royalties and to report on orders, date of shipment, amounts, sizes, types, and prices. Also Aerofin consulted these manufacturers prior to its issue of price lists. A consent decree, with the same date as the complaint, cancelled all patent licenses based on this price-fixing program and provided royalty-free licenses to any prospective manufacturer of either encased or uncased coils.[67] Aerofin and the other defendants engaged in these alleged practices from 1928 to 1943.

b. *Cast-Iron Pressure Pipe.* U. S. Pipe and Foundry, founded in 1899 as a consolidation of the old Addyston Pipe and Steel Company and other companies, became in 1921 the exclusive licensee in the United States for the centrifugal method of making cast-iron pipes. Between 1922 and 1946, it entered into lease-license agreements with four other companies under which it leased machines for producing centrifugal pipe and granted licenses to use the patents covering them. A government complaint of 1947 charged that U. S. Pipe and Foundry limited the output of such pipe and set the prices. The licensees agreed to use these machines only for making such pipe, to employ no rival machines, to assign any kindred patents to the licensor, and to confine their sales to the United States. A consent decree in 1948 required the discontinuance of these restrictions, the transfer of title to pipe-manufacturing machines, and the dedication of fifty patents to the public.[68]

A similar situation and outcome involved the Sand Spun Patents Corporation and three other companies. Sand Spun, the owner or exclusive licensee of nineteen patents concerning the centrifugal method of making pipes, granted licenses under these patents on conditions which sought to perpetuate its control of patents, production of pipe,

[67] *United States* v. *Aerofin Corp.*, CCH Trade Regulation Reports, Supp. 1941–43 ¶ 52,913 (S.D.N.Y. Civil No. 20–458; Complaint and Consent Decree, Mar. 5, 1943).

[68] *United States* v. *United States Pipe and Foundry Co.*, CCH 1948–49 Trade Cases ¶ 62,285 (D.N.J. Civil No. 10772; Complaint, 1947; Consent Decree, 1948).

and allocation of markets. A consent judgment of 1949 dissolved Sand Spun, enjoined the restrictive agreements, and dedicated its nineteen patents to the public.[69]

c. *Stainless Steel.* Eighteen leading stainless steel companies fixed prices by means of patent licensing, and a consent judgment in 1948 enjoined the practice and required that licenses be granted to any new applicant at reasonable royalties.[70]

d. *Screws and Drivers.* According to a complaint of 1947, the Phillips Screw Company, from 1937 to 1947, sublicensed manufacturers to make and sell cross-recessed screws and screw drivers under price and other restrictions and entered into territorial agreements with foreign producers. Sales were largely to automotive and airplane manufacturers and to distributors. In 1947, there were thirty-three manufacturers of this kind of screw and only four of them were "independents." A consent decree of 1949 enjoined license agreements for fixing prices and dividing either domestic or international markets and required licenses be granted under twenty-eight patents at reasonable royalties and with technical information.[71]

e. *Alloys and Dentures.* Austenal Laboratories owned or controlled certain patents covering the casting of dentures from alloys and the refractory molds for fabricating the dentures. Their licensed dental laboratories, it was charged, agreed to purchase all requirements of alloys from Austenal, to use the alloys for dental purposes only, to sell finished dentures at prices fixed by this company, and to limit sales of dentures to specified areas. Austenal used Vitallium as a trademark for both its alloys and dentures and permitted the licensed laboratories to advertise it only in connection with dentures made from its alloys. A consent decree of 1951 granted patent and trademark relief and enjoined these restrictive practices.[72]

f. *Daylight Fluorescent Materials and Products.* According to a

[69] *United States* v. *Sand Spun Patents Corp.,* CCH 1948–49 Trade Cases ¶ 62,462 (D.N.J. Civil No. 125–49; Complaint, 1949; Consent Judgment, 1949).

[70] *United States* v. *Allegheny Ludlum Steel Corp.,* CCH 1948–49 Trade Cases ¶ 62,330 (D.N.J. Civil No. 45–83; Complaint, 1945; Consent Judgment, 1948).

[71] *United States* v. *Phillips Screw Co.,* CCH 1948–49 Trade Cases ¶ 62,394 (N.D. Ill. Civil No. 47–C–147; Complaint, 1947; Consent Judgment, 1949).

[72] *United States* v. *Austenal Laboratories,* CCH 1950–51 Trade Cases ¶ 62,880 (S.D.N.Y. Civil No. 50–496; Complaint, 1949; Consent Decree, 1951). A similar case was *United States* v. *Consolidated Car-Heating Co.,* CCH 1950–51 Trade Cases ¶ 62,656 (S.D.N.Y. Civil No. 34–312; Complaint, 1945; Consent Judgment, 1950).

complaint, Switzer Brothers entered into license agreements directly or through other companies with about 1,200 purchasers and users of daylight fluorescent materials which allocated products and customers and imposed other unlawful restrictions. Consent judgments in 1953 provided for royalty-free licensing under three patents and compulsory licensing with reasonable royalties under six others.[73]

g. *Fluorescent Lamps*. General Electric, along with Westinghouse, had many patents in the fluorescent lamp field. According to a complaint of 1942, restrictive practices enabled Corning to become virtually the only manufacturer who sold clear glass tubing and the only source of machinery for making it; GE became the only source of sale of coated glass tubing to manufacturers of neon signs and was the only manufacturer and seller of fluorescent lamp boxes. A predominant proportion of fluorescent lamps, first placed on the market in 1935, was made and sold by GE, Westinghouse, and Consolidated. Westinghouse and Consolidated had license agreements which limited their output to certain amounts. Under a contract of 1927, Westinghouse agreed to sell fluorescent lamps at the same prices and upon the same terms fixed and followed by GE.

This complaint, after amendment and postponement because of war, led to a consent decree against Corning in 1946. This company was required to grant licenses royalty-free under its present patents and at reasonable royalties under its future patents applied for before 1940; to furnish at a nominal charge technical information on glass formulas in use on date of judgment; to sell glass bulbs, tubing, and cane at nondiscriminatory prices; and to refrain from dividing the market and excluding others and from lessening competition by acquiring any domestic company.[74]

Final consent judgments against other companies were not entered until 1954. One of them enjoined General Electric from entering into any arrangement with any manufacturer of lamps, lamp parts, and lamp machinery to fix prices, allocate markets, limit production, or restrict imports and exports. Moreover, it provided for compulsory royalty-free licensing of the patents of this company covering these

[73] *United States v. Switzer Brothers, Inc.*, CCH 1952–53 Trade Case ¶67,605 (N.D. Calif. Civil No. 29860; Complaint, 1950; Consent Judgments, 1953).

[74] *United States* v. *General Electric Co., Corning Glass Works, et al.*, CCH 1946–47 Trade Cases ¶ 57,448 (D.N.J. Civil No. 2590; Complaint, 1942; Consent Decree *re* Corning, 1946).

products. Separate judgments required Westinghouse and Consolidated also to refrain from restrictive agreements. Furthermore, the Westinghouse judgment required the royalty-free licensing of the fluorescent patents of this company.[75]

h. *Milling Machines*. A complaint of 1954 charged three companies with license agreements by which they divided manufacturing fields among themselves, refused licenses to others except upon restrictive terms, refrained from patent infringement suits against each other, and required similar immunity from their assignees and licensees. A consent decree of the same date enjoined these restrictive practices, required these companies to grant licenses on a reasonable royalty basis and to provide specifications and drawings for five years upon application of any licensee.[76]

Complaints. Recent complaints of the Department of Justice indicate the following instances of license agreements with price and other restraints among former competitors: First, one company became the exclusive licensee under a British-owned patent on alignate dental impression powder and limited sublicenses to two other companies, one of which had its own patents in this field but refused to grant licenses; this arrangement prevented competition between these companies and alternate inventions.[77] Another complaint alleged restraint of trade among six manufacturers of sheet chargers used in feeding sheets of metal materials to rolling mills in the metals machinery industry.[78] Third, in 1933, Browne and Sharpe Manufacturing Company and others organized Associated Patents for holding and licensing their present and future patents pertaining to machine tools; by this means they allocated among themselves the manufacture and sale of such products according to specified types and on an exclusive basis.[79] Lastly, two manufacturers controlled competition between themselves and with others in the manufacture and sale of induction-hardened crank shafts by allocating customers.[80]

[75] *Ibid.*, CCH 1953–54 Trade Cases ¶67,794–67,796 (D.N.J. Civil No. 2590; Complaint, 1942; Consent Decree *re* General Electric, *et al.*, 1954).

[76] *United States* v. *Cincinnati Milling Machine Co.*, CCH 1953–54 Trade Cases ¶ 67,733 (E.D. Mich Civil No. 13401; Complaint and Consent Judgment, April 19, 1954).

[77] *United States* v. *L. D. Caulk* Co. (D. Del. Civil No. 1372; Complaint, 1951).

[78] *United States* v. *The Logan Co.* (W.D. Pa. Civil No. 96–58; Complaint, 1951).

[79] *United States* v. *Associated Patents, Inc.*, (E.D. Mich. Civil No. 10–664; Complaint, 1951).

[80] *United States* v. *Ohio Crankshaft Co.* (N.D. Ohio Civil No. 28299; Complaint, 1951).

Resale Price Maintenance

Resale price maintenance is the dictation by the seller of the actual or minimum prices at which buyers resell the product purchased from him. The license agreements of the manufacturer with distributors and some of those with other manufacturers have sought this objective. Control over price through bona fide agents or by any other means that do not require a resale is not of course maintenance of resale prices. This price policy is confined ordinarily to trade-marked articles which reveal the producers of them. The General Electric decision of 1926 indicated that manufacturing licensees may be legally required to observe certain prices. May a patentee compel purchasing licensees to adhere to certain prices? In other words, may a patent-owning company legally maintain the resale price of the distributor of its products?

Attempts along this line have generally been through either a license agreement with a distributor or a so-called printed notice on the product. The Supreme Court in 1911 held that a manufacturer could not legally establish the resale prices of its trade-marked product through a system of contracts and notices.[81] But within recent years, forty-five states have adopted laws which legalize such contracts. Each of these laws, of course, applies only to commerce within a particular state. In 1939, Congress passed the Miller-Tydings Act which approves similar legislation for commerce between these states; specifically it permits a manufacturer of trade-marked goods to agree with his distributors as to resale prices. An examination of several court cases, however, will point out some of the legal questions that have been involved in the methods of price maintenance.

License Agreements. a. Electric Lamps. In 1911, the government charged General Electric and numerous subsidiaries with violation of the Sherman Act by price fixing. A circuit court, though recognizing the right of these patent-owning companies to grant licenses upon terms and conditions, prohibited them "from requiring or imposing upon the licensee the fixing of a resale price to be observed by his vendees." The court declared that the purchaser of lamps "whether at wholesale or retail, shall not be in any manner restricted as to the price at which such lamps shall be sold to the public or to

[81] *Dr. Miles Medical Co.* v. *Park and Sons Co.,* 220 U.S. 373 (1911).

any dealer or consumer."[82] Note that this decree left the way open for GE to fix the prices of lamps of manufacturing licensees—a fact which may have influenced the Supreme Court in its approval of the practice in 1926.

In 1912, however, GE achieved retail price control in another way. GE entered into contracts with four hundred wholesalers and twenty-one thousand retailers by which these dealers became its agents to sell on commission. It consigned the lamps to them, transportation prepaid, and fixed the prices at which they sold the lamps. Dealers paid the expenses of all other functions—storing, handling, selling, delivering to customers, and assuming risks of breakage and losses from uncollectible accounts. Not later than the fifteenth of each month dealers paid an amount equal to the total sales value less commission on all lamps sold. GE assumed all risks of fire, flood, obsolescence, and price decline, and paid taxes assessed on the lamps.

In its action against GE in 1926, one phase of which is described earlier in this chapter, the government alleged that GE's system of marketing was merely a device to fix the resale prices of lamps in the hands of purchasers. In view of this charge, the issue, according to the court, was whether these dealers "are to be treated as agents, or as owners of the lamps consigned to them under such contracts."[83] The Supreme Court decided that they were "genuine agents"—that they received consignments for their custody and sale. Its conclusion was, "The owner of an article, patented or otherwise, is not violating the common law, or the antitrust law, by seeking to dispose of his article directly to the consumer and fixing the price by which his agents transfer the title from him directly to such consumer."[84]

The principal characteristic of a commission man in selling goods on consignment—the determination of the selling price independent of his principal—was missing in this manufacturer-dealer relationship. Also in consignment of merchandise, commission, interest, and other expenses of selling, financing, and storing, are generally charged to the owner. Actually GE sold lamps on thirty-day credit to dealers, stipulated the retail price of their lamps, and fixed their gross margin of profit. With words rather than deeds it sold on consignment.

[82] *United States* v. *General Electric Co.,* I D & J. 267, 270 (1911).
[83] *Ibid.,* 272 U.S. 476, 484 (1926).
[84] *Ibid.,* 488.

In more recent times, GE and Westinghouse have used agency agreements with distributors to maintain resale prices on fluorescent as well as incandescent lamps distributed by wholesale and retail "agents." In 1949, Judge Forman held that these agreements did not violate antitrust laws but that the manner in which the operation of their two agency systems was enmeshed through the office of a GE supervisor was unlawful. After 1933, when its basic patents had expired, "General Electric was without authority to compel Westinghouse to submit for its approval, as to responsibility, candidates for agency appointments."[85]

Carboloy Company, a GE subsidiary, controlled resale prices of tungsten carbide through so-called agents. Judge John C. Knox, in 1948, said "In my judgment, the evidence justifies a finding that the arrangement violated the prohibition against resale price fixing for two reasons: (a) the agents were really purchasers, and (b) the agents were competitors within the meaning of *United States* v. *Masonite Corp.*"[86]

b. Lenses. The Univis Lens Company owned a patent on a method of making lens blanks and of grinding and polishing them and thus converting them into finished lenses for use as eyeglasses. It created the Univis Corporation, a dummy, to which this and other patents were transferred. The so-called corporation licensed Univis Lens to manufacture lens blanks, to sell them to its designated licensees, and to receive a royalty of fifty cents a pair. Univis Lens had three classes of licensees—wholesalers, prescription retailers, and finishing retailers. The wholesalers were authorized to purchase the blanks from Univis Lens, to finish them by grinding and polishing, and to sell them to prescription retailer licensees. These retailers, about one-half the total number, agreed in turn to sell the finished lenses only to consumers. The finishing retailers, who polished and ground and adjusted the lenses to the eyes of the consumers, had licenses. The wholesalers and retailers paid no royalties. Any reward that Univis Lens and its dummy received from the patent was derived entirely from the sale of lens blanks. In all cases the dummy corporation fixed the prices for each pair of lens: $3.25 from manufacturer to wholesaler; $7 from wholesaler to retailer; and $16 for the white and $20 for the

[85] *Ibid.*, 82 F. Supp. 753, 866 (D.N.J. 1949).
[86] *Ibid.*, 80 F. Supp. 989, 1006 (S.D.N.Y. 1948).

tinted, from the prescription retailer to the consumer. The finishing retailer paid $4 a pair, and like the prescription retailer, received $16 and $20. Deviation from these fixed prices resulted in cancellation of licenses.[87]

The Supreme Court concluded that "merely because the licensee takes the final step in the manufacture of the patented product, by doing work on the blank which he has purchased from the patentee's licensee, it does not follow that the patentee can control the price at which the finished lens is sold."[88] The court held that "an incident to the purchase of any article, whether patented or unpatented, is the right to use and sell it, and upon familiar principles the authorized sale of an article which is capable of use only in practicing the patent is a relinquishment of the patent monopoly with respect to the article sold."[89] Therefore, "the patentee cannot control the resale price of patented articles which he has sold, either by resort to an infringement suit, or, consistently with the Sherman Act (unless the Miller-Tydings Act applies), by stipulating for price maintenance by his vendees."[90] And the court pointed out that even the Miller-Tydings Act could not apply in this situation because Univis Lens manufactured the blanks and not the finished lenses to which the resale prices applied. It did not make the product which the licensees sold, and the licensees therefore were not engaged in the "resale" of the same commodity they bought. "We find nothing in the language of the Miller-Tydings Act, or in its legislative history, to indicate that its provisions were to be so applied to products manufactured in successive stages by different processors that the first would be free to control the price of his successors. The prescribed prices are thus not within the Miller-Tydings exception to the Sherman Act."[91]

The court therefore declared unlawful the entire licensing and price-fixing system other than the Univis Corporation's license to Univis Lens.

In another case involving optical goods, Bausch and Lomb and Soft-Lite Lens were charged with maintaining uniform resale prices of their products. The Supreme Court, in 1944, said, "Even the addi-

[87] *United States* v. *Univis Lens Co.,* 316 U.S. 241, 243–45 (1942).
[88] *Ibid.,* 249.
[89] *Ibid.,* 249.
[90] *Ibid.,* 250.
[91] *Ibid.,* 253–54.

tional protection of a copyright, or of a patent, adds nothing to a distributor's power to control prices of resale by a purchaser."[92] However, this interpretation should not be confused with legislation in forty-five states that specifically permits price maintenance for trade-marked articles, patented or unpatented.

c. Other Products. Various patent pools and consolidations already described employed patents for the maintenance of resale prices. Their dominance derived from patents lent itself effectively to such a policy. A few illustrations may be cited briefly. American Optical and other optical companies prescribed the prices of wholesaler and retailer for frames and lenses. Further, Bendix Aviation and Wagner Electric appointed distributors of their braking systems under arrangements which gave them control over resale prices and policies. Again, the agreement between The American and Phillip Screw companies and their sublicensees provided for the maintenance of resale prices which were determined by all of them on the basis of their costs and opinions.

Printed Notice. Another phase of license agreements for resale price maintenance is the so-called printed notice. In many instances the manufacturer attached a notice to his product which stated that the dealer should not resell it at less than a stipulated amount and that any purchase or sale for less constituted infringement. At first the lower courts upheld the practice; one of the decisions was written by Judge Willis Van Devanter, who later became a member of the highest tribunal.[93]

In 1908, the Supreme Court held that the owner of a copyright on a book did not have the legal right under the copyright law to fix by notice the resale prices of copies of the book. This was its first decision definitely denying the right to maintain resale prices.[94] A similar opinion was handed down in 1913.[95] There is a basic similarity in issues under patent and copyright laws. The Supreme Court has always treated them as the same, using cases in the two fields interchangeably as authority.

a. Sanatogen. Bauer and Company made and sold a patented drug

[92] *United States* v. *Bausch and Lomb Optical Co.*, 321 U.S. 707, 721 (1944).

[93] *National Phonograph Co.* v. *Schlegel,* 128 Fed. 733 (8th Cir. 1904).

[94] *Bobbs-Merrill Co.* v. *Straus,* 210 U.S. 339 (1908).

[95] *Straus* v. *American Publishers' Association,* 231 U.S. 222 (1913).

called Sanatogen. Each container bore a "Notice to the Retailer," as follows:

> This size package of Sanatogen is licensed by us for sale and use at a price not less than one dollar. Any sale in violation of this condition, or use when so sold, will constitute an infringement of our patent No. 601,995, under which Sanatogen is manufactured, and all persons so selling or using packages or contents will be liable to injunction and damages. A purchase is an acceptance of this condition. All rights revert to the under-signed in the event of violation. The Bauer Chemical Company.[96]

Accordingly, when O'Donnell, a druggist, advertised and sold original packages of Sanatogen at less than the prescribed price, Bauer brought suit for infringement. The Supreme Court stated the issue as follows: "May a patentee by notice limit the price at which future retail sales of the patented article may be made, such article being in the hands of a retailer by purchase from a jobber who has paid . . . the full price asked for the article sold?"[97] In a five to four division it gave a negative answer. "The packages were sold with as full and complete title as any article could have been sold in the open market, excepting only the attempt to limit the sale or use when sold for not less than one dollar. In other words, the title transferred was full and complete, with an attempt to reserve the right to fix the price at which subsequent sales could be made."[98] Moreover, "There was no transfer of a limited right to use this invention, and to call the sale a license to use is a mere play upon words."[99]

What is the distinction between this opinion and the decisions of the lower courts relating to this same issue? The lower courts stated that the patentee, who has the exclusive right to vend his product, agreed to part with it if the purchasers would maintain the resale price, and that the sale was conditional and incomplete. The Supreme Court held that the patentee received full consideration for his commodity, that the title passed to the dealer, and that the sale was unconditional and complete.

b. *Talking Machines.* In 1917, the Supreme Court reaffirmed the rule of the Sanatogen case by denying to Victor Talking Machine

[96] *Bauer* v. *O'Donnell,* 229 U.S. 1, 8–9 (1913).
[97] *Ibid.,* 11.
[98] *Ibid.,* 16.
[99] *Ibid.,* 16.

Company the right to fix resale prices. The Victor company employed a "License Notice" attached to its machines and a "License Contract" with dealers in disposing of its machines. Both of them referred to "license to use" and "royalty" rather than price; specified that "title to the machine shall remain in the Victor company"; reserved the right to the manufacturer "to inspect, test, and repair the machine at all times, and to instruct the user in its use," without however his assuming any obligation to do so; and provided that the so-called licensee shall become the owner upon the expiration of the patent having the longest time to run. According to the court, the sole act of alleged infringement by the defendant was the purchase and sale of Victor machines for less than the prices stated in the notice.[100]

"It thus becomes clear," the court stated, "that this 'License Notice' is not intended as a security for any further payment upon the machine, for the full price, called a 'royalty,' was paid before the plaintiff parted with the possession of it; that it is not to be used as a basis for tracing and keeping the plaintiff informed as to the condition or use of the machine, for no report of any character is required from the 'ultimate user' after he has paid the stipulated price; that, notwithstanding its apparently studied avoidance of the use of the word 'sale' and its frequent reference to the word 'use,' the most obvious requirements for securing a bona fide enforcement of the restrictions of the notice as to 'use' are omitted; and that, even by its own terms, the title to the machines ultimately vests in the 'ultimate users'; without further payment or action on their part, except patiently waiting for patents to expire on inventions, which, so far as this notice shows, may or may not be incorporated in the machine."[101] The court concluded that the purpose of the notice, as in the Sanatogen case, was to restrict the price of the machines after the manufacturer had been paid for them and after they had passed into the possession of dealers and the public.[102] Justices Joseph McKenna, Oliver Wendell Holmes, and Van Devanter dissented.

The American Graphophone Company sold its patented products to the dealer with the stipulation in the contract that the products must be resold at prices fixed by this company. It brought suit against the

100 *Straus* v. *Victor Talking Machine Co.*, 243 U.S. 490, 494–95 (1917).
101 *Ibid.*, 500.
102 *Ibid.*, 501.

Boston Store for violation of this contract. American Graphophone argued that the wording of its contract and notice was important, that it differentiated its scheme of price maintenance from other attempts of patent owners to dictate resale prices. In meeting this argument the Supreme Court stated, "It becomes, we think, unnecessary to do more than say that we are of the opinion that the attempt in argument to distinguish the cases by the assumption that they rested upon a mere question of the form of notice, on the patented article, or the right to contract solely by reference to such notice, is devoid of merit, since the argument disregards the fundamental ground upon which, as we have seen, the decided cases must rest."[103] Therefore the court gave a negative answer to the definite question, "Can a patentee, in connection with the act of delivering his patented article to another for a gross consideration then received, lawfully reserve by contract a part of his monopoly right to sell?"[104]

Conclusion

In the incandescent lamp case of 1926 the Supreme Court emphasized the right of a patent owner to license manufacturers with restrictions as to price. In the Ethyl, Masonite, Gypsum, and other cases of later years, the court inquired into the effect of these restrictions, particularly upon former competitors. In several instances stricter standards were imposed in reference to multiple licensees. The right of a patent owner to require licensed manufacturers to adopt its prices and plan of distribution as announced in 1926 in the GE case was seriously weakened by a four to four vote of the Supreme Court on this question in the Line Material case of 1948. A few companies accepted consent decrees requiring either dedication of patents or royalty-free licenses even after the Hartford-Empire decision of 1945 which disapproved compulsory licensing without payments. The final judgment of a district court in 1953 provided that GE and other defendants must dedicate to the public their patents on incandescent lamps and lamp parts—the first court order of its kind in antitrust litigation.

Whether a patent owner can legally employ so-called license agree-

103 *Boston Store* v. *American Graphophone Co.,* 246 U.S. 8, 25 (1918).
104 *Ibid.,* 27.

ments to dictate the prices at which distributors sell an invention depends upon whether they are actually licensees or bona fide agents on the one hand or purchasers of the products on the other. The licensee or agent may be legally bound to sell at prescribed prices by the patent owner. However, the purchaser of a product, whether patented or unpatented, has full control of the right to sell it. In other words, a patentee cannot control the price of his patented product with respect to persons purchasing it either from him or his licensee.

The Supreme Court in several decisions, the first one in 1913, denied the right to dictate the resale prices of patented goods by printed notice. It held that the same legal principles apply in the marketing of both patented and unpatented articles.

In the General Electric case of 1926 the Supreme Court approved price restrictions upon the manufacturing licensee but its decisions in the Sanatogen, talking machine, and kindred cases consistently condemned any attempt to control the price of a purchasing "licensee." One explanation lies in the contracts of the two so-called licensees. One of them, a manufacturer, was a business enterprise and a party to a written contract while the other was usually one of many retail merchants or one of countless consumers. The users, and even retailers in some instances, did not negotiate and enter into agreements covering price or anything else. Any effort of the patentee to restrict them was arbitrary or dictatorial—a printed notice in the box of merchandise or on the product or some other impersonal device.

The Line Material case of 1948, in which the GE decision of 1926 was undermined though not discarded, returned to the basic issue—does the patent give its owner the right to control the prices of the manufacturers, users, and sellers or purchasers of its invention? Does the patent owner have the exclusive right to make, use, and sell the invention or exclude others from doing so—and nothing more? May the owner also legally extend this right to manufacturing and marketing and using by others and thus control their production, sales, and uses? In other words, is the patented product completely immune from antitrust legislation?

CHAPTER V

Cartels

THE WORD "CARTEL," as used in this book, means a monopoly or restraint of trade on an international basis. Whenever possible, patents are used as a means to achieve effective control and the semblance of legality. Patent cartels may restrict market territories, fields of operation, customers, output, use of others' patents, prices, purchases, and suits against validity of patents. Some of them provide for an exchange of patent rights, both present and future, and of confidential information essential to the practical working of the inventions. Patent cartels are found in many fields, particularly in the chemical and electrical, and the members are usually located in the great industrial nations.

Whether a patent cartel is within the law depends upon political and economic policies and therefore upon laws and their interpretation. Except within recent years the cartel has had its own way. The first one exposed and condemned by the Supreme Court was in 1947. Its legal status, like that of monopoly and restraint of trade within the United States, depends upon various factors.

Types

As just suggested, the use of patents for the control of the market among nations, like that within a nation, may be achieved by pools, consolidations, and license agreements. A cartel may involve two or more of these types; for example, each member that enters into a pool may represent a consolidation of patents in its own country and may grant licenses of its own outside the consolidation.

Pools. In pooling or exchanging patents on a world-wide basis in order to restrain trade, each member is usually given the exclusive right to the patents of other members as well as its own issued by its home country. The member thus obtains control of the market in his own country and in its possessions. Other parts of the world, usually less developed countries in South America and Asia, are either divided among the members or thrown open to all. Pooling kindred patents of dominant companies in all nations, rather than in one country only, permits a more exclusive and arbitrary control. If outsiders become licensees, they must accept dictatorial terms as to validity of patents, assignment of their own inventions, and so forth. Apparently there is no instance of letting members and outsiders use the pooled inventions on a royalty basis and sell their products where and as they please—of letting them operate as members did in the gasoline-cracking pool upheld by the Supreme Court in 1931.

A variation of the pooling scheme is for United States corporations to grant exclusive licenses to foreign companies but to receive in return nonexclusive grants in order to avoid the appearance of monopoly or restraint of trade in the United States.

Consolidations. One company may become dominant not only in one country but in the whole world. The control of certain dyes and chemicals by I. G. Farbenindustrie Aktiengesellschaft of Germany—generally known in this country as "I. G." or "Farben" or "Farbenindustrie"—affords an excellent illustration. Primarily through its own research but also through purchases it acquired patents on practically all actual and prospective inventions in specific fields. Again, two or more companies may become dominant in their respective territories from the cross-assignment of patents and patent applications. As explained later, Farben and Standard of New Jersey divided control of certain commodities. Such a monopoly may exploit the patents taken out in any foreign country by reserving that country as the market for the factory at home, by assigning or licensing the patents to a subsidiary located there, or by collecting royalties from licensees. As just indicated, the manufacture may take place either in the foreign market or the homeland. The ideal situation is to reserve foreign markets through patents and to send goods to those markets with little or no hindrance from tariffs. In any case, there is an effort to continue the monopoly by obtaining improvement patents and by retaining the know-how for the most advantageous use of the invention. James E. Markham, alien property custodian, stated in 1945 that General Aniline and Film Corporation was dependent upon Farben for important research and know-how. "The result of this policy was the complete subservience of General Aniline to its German associate, for the results of the German research were never fully disclosed to the company. . . . And in many cases important material was only communicated verbally to the most trusted employees of the company on the occasion of their visits to Germany for use as I. G. Farbenindustrie might direct. The information thus obtained was not disclosed to other employees of the company. Thus, on several occasions, when the man in possession of the information died, General Aniline was obliged to send another employee to Germany for instruction in the particular process."[1] The strategy of course was to keep the monopoly even if the physical assets and patents were ever seized by the United States—as they actually were in 1942.

License Agreements. As already noted, different owners of competing patents may pool or cross-license their patents and divide the markets. Another possibility is for one of the owners, usually the one in the strongest position, to grant licenses to the others upon restrictive conditions as to territory, output, and prices. The result is that each of them has a monopoly in a particular market. Another arrangement is for these owners to divide the world among themselves and then for each of them to grant restrictive licenses within its own area. These schemes, like the pool, permit each member to become dominant in a particular region. In some instances United States corporations participate in such schemes by understanding rather than by license.

Reasons for Growth

Various factors have contributed to the growth of cartels. The chief one is a definite program to exploit patent systems in order to accomplish certain economic and political objectives. During the twentieth century important nations have been deprived of all or parts of their empires; note the loss of colonies by Germany and Italy and Japan after military defeats, and the complete or partial liberation of such dependencies as Cuba and India. Because of these losses, industries within these nations turned to new schemes for getting and keeping markets for their goods. They encouraged research nearby rather than exploration far away. They reserved their markets in the name of patents rather than the flag, and thus depended upon patent laws

[1] EGRW Hearings, Part 4, 590.

and agreements rather than conquest. Territory was divided, not by pope or treaty, but through cartel agreements by business corporations. In effect, one kind of colonization supplanted another, and countries politically free became economic dependencies.

Germany was the leader in this movement. It subsidized research and encouraged the securement of patents abroad as well as at home and the incomplete disclosure of inventions. It permitted the concentrated control of sales at home and abroad and maintained a high tariff to keep out foreign products. Germany and its business leaders worked together in promoting the interests of both. In later years they became affiliated together in Farben which in turn became dominant in many cartels. Farben was a conglomerate concentration of about four hundred enterprises in Germany with factories, power plants, and mines, and had financial interests in a similar number of enterprises abroad. Its general field of dominance was the chemicals. Before World War I, German companies took out patents in other countries and in effect reserved them as markets, especially for chemicals and dyes. Later this practice encountered new patent laws in Great Britain and elsewhere, though not in the United States, which sought to prevent the suppression of patents, and encountered higher tariffs, especially in the United States in 1930 and in Great Britain in 1932. To avoid these obstacles Farben promoted international agreements with industrial leaders in other nations that made each industry exclusive in its own country but that divided the rest of the world among the cartel members.

Scientific advance, especially in the chemical and electrical fields, increased the need for exchanging technology between nations as well as within a nation. Again, the world depression in the thirties was the background for desperate measures, public and private, to check the decline in prices. Also by that time, in various industries, dominance or leadership had been established—for example, in the chemical industry, Farben dominated in Germany, Imperial Chemical Industries in England, and Du Pont in the United States. Further facilitating the patent cartel was the concentration of patents as well as physical assets in each industrial nation. There can be no monopoly in the whole world prior to monopoly in each of its parts. By the same token, it is necessary to continue the local monopoly or else the "exclusive market" awarded the local member by other cartel members becomes farcical. In the United States at least, a lower tariff, attained through a reciprocity program beginning in 1934, made it more necessary than ever for electrical, chemical, and some other manufacturers to organize or join cartels in order to eliminate competition from abroad as had already been done at home. As stated in 1950 by Winthrop G. Brown of the Department of State, "It has been found that as governmental barriers to international trade have been broken down through reductions in tariffs and similar measures, barriers created by agreement among producers have tended to rise to replace them."[2] A cartel, like a tariff, restricts the movement of goods between nations. An actual or prospective member of a cartel derives extra bargaining power from behind a high tariff because the member is already assured of its home market. Lastly, the successful monopoly or restraint of trade at home through the concentration of patents, such as occurred in the United States and other countries, suggests the same result and methods on a world-wide scale.

Objectives

Economic. The immediate objective of course is to get patents and confidential information; for example, the agreement between Du Pont and Imperial Chemical Industries in 1929 admitted that "each of the parties hereto desires to acquire licenses in respect of the patented and secret inventions of the other party."[3] What are the ultimate objectives? Alleged economies from regularity, certainty, and concentration—more specifically, avoidance of cross freights and duplicated efforts in selling, and so forth—are always advanced by those who defend totalitarianism, whether socialism or communism in government or monopoly in private enterprise. Such economies at their best are very limited in amount and duration and at their worst block economic progress. Competitive wastes are an unavoidable price for technological advance. How can the best of alternate inventions emerge except through rivalry in making, using, and selling?

The dominant objective in a patent cartel is control of the market —a control that permits economic gains and, in some instances, political power in peace and military strength in war. Control of foreign

[2] M P Hearings, Part 3, 147–48.
[3] Bone Hearings, Part 5, 2278.

markets may be exercised as effectively through patents and secret inventions as through a colonial relationship. By concentrating patents and know-how, one or more companies may in effect reserve markets for themselves. Backward areas are divided up somewhat as the new world was divided up centuries earlier except that companies with patents and know-how replace nations with armies and navies. In backward areas foreign companies build up favorable balances of trade with which to obtain their raw materials. They may do this by selling merchandise made at the home factory or in a branch factory abroad or by collecting royalties from foreign licensees. Also, foreign companies may compel backward areas to buy other supplies and services, just as colonies once had to receive and send goods only in the ships of the mother country. The most highly-prized situation of course is the single company with sufficient dominance in a field to reserve the entire world as its market. As set forth by a Department of State official in 1945, "The great German combines were the spearheads of Axis economic penetration in the other American republics. In the field of drugs and pharmaceuticals the Bayer, Merck, and Schering companies enjoyed a virtual monopoly. I. G. Farben subsidiaries had a firm hold on the dye and chemical market."[4] Such an empire is better than colonies because the upkeep is limited to the expense of obtaining and maintaining patents and retaining secret know-how.

Cartel members generally prefer not to establish branch factories in small backward countries, and therefore they oppose either patent laws or tariffs which seek to encourage industries within such countries. And they are ever alert, of course, in seeking greater protection of patent rights by pressing for adoption of stronger patent laws and stronger enforcement procedures by national governments throughout the world and by seeking maximum recognition of patent rights through such international agreements as the International Convention for the Protection of Industrial Property.

Other economic objectives besides control of market include avoidance of litigation and acceptance of validity of patents. Acceptance of membership in a cartel means avoidance of litigation with foreign companies. According to a report of a meeting in 1935 between Du Pont and Farben representatives, one phase of co-operation in the

[4] EGRW Hearings, Part 2, 62.

field of dye stuffs consists of "amicable settlements of all patent litigation in the United States." This statement applies also of course to interferences, either actual or prospective. Also, cartel membership means, by agreement, that there will be no challenge of validity of present or future patents from other members. Such an assurance eliminates most of the possible tests of validity because outsiders usually lack both the interest and the financial means to take such action.

Political. Cartels, especially the ones based on patents, provide a means for the "effective division of national spheres of influence."[5] The cartel apportions countries as markets among its members just as treaties provide mandates and spheres of influence among nations. In some cartels United States corporations are restricted to the home market and foreign companies are awarded Central and South America. The effect of this, to the extent that political influence results from economic control, is to reverse the Monroe Doctrine—keeping the United States out of and letting other nations enter the American countries to the south. Regardless of intentions, any nation which is the sole source of supply to another for essential products and services, such as dyes and transportation, may assume undue significance. The flag may even follow the trade.

Germany made the most of its opportunity in Central and South America. "When South Americans sought to purchase drugs, metals, precision equipment, and munitions from the United States, private cartel treaties had already provided that American concerns could not engage in this trade. Not only was the healthy development of South American trade and industry checked, but even today we struggle desperately to overcome the political consequences of cartel activity."[6] A dramatic illustration of political influence is that Farben's legal counsel in France became the betrayer of that country during the recent war.

Cartel agreements have the effect of treaties between invisible but supergovernments. "These cartels have established private treaties—without government participation—between domestic and foreign monopoly groups to fix prices, divide market territories, and fields of production."[7] They are in effect private tariff agreements to re-

[5] Bone Hearings, Part 1, 648.
[6] Wendell Berge, *Cartels,* 12.
[7] CSB Report, 93.

strict the movement of goods. Senator Homer T. Bone stated in 1943 that cartel agreements were more "effective in shaping the lives and destiny of the people of this country than we accomplished by many treaties which were formally ratified by the United States Senate."[8] According to the Department of Justice, "International cartels establish their own private tariff walls and export and import restrictions, in many cases frustrating the policy of the United States as well as destroying free competition in foreign and domestic commerce."[9] That members of a cartel may regard themselves as supergovernments, only temporarily inconvenienced by national governments, is indicated in a report made by a Standard of New Jersey official October 12, 1939: "Pursuant to these arrangements I was able to keep my appointments in Holland where I had three days of discussion with the representatives of the I. G. They delivered to me assignments of some two thousand foreign patents, and we did our best to work out complete plans for a *modus vivendi* which could operate through the term of the war, whether or not the United States came in."[10]

Germany, in particular, used patent cartels to accomplish military objectives. "The cartel program as developed first by the General Staff and the big industrialists, and after 1933 by the Nazis as well, called for the manipulation of the world's trade and the world's resources by German industry through the medium of cartel arrangements so as to strengthen Germany's position to make war, and, in turn to weaken the defensive position of its potential enemies."[11] Germany's cartel program hurt, of course, the United States. "On numerous occasions in the United States, German companies through patent agreements were able to obtain highly confidential technical data."[12] According to a Farben member, the Nazi government used German industry as one instrument "to weaken the military strength of the other countries and to keep the Wehrmacht supreme."[13] In 1942, a report to President Roosevelt declared that the enemy "has worked for many years to weaken our military potential. Through patent controls and cartel agreements he succeeded in limiting Ameri-

[8] STM Hearings, Part 4, 321.
[9] CSB Report, 251.
[10] Bone Hearings, Part 6, 2639.
[11] EGRW Hearings, Part 3, 177.
[12] *Ibid.*, 460.
[13] *Ibid.*, Part 10, 1289.

can production and export of many vital materials. He kept the prices of these materials up and the output down. He was waging war, and he did his work well, decoying American companies into agreements the purpose of which they did not sense."[14] Illustrations, detailed later, include magnesium, synthetic rubber, dyes, pharmaceuticals, and optical instruments.

A cartel member affiliated with its government, illustrated by Krupp and Farben of Germany, had a distinct advantage over a member that acted alone. It generally had greater bargaining power for several reasons. In some instances it had more to trade because of government subsidy or research. It represented the entire industry, as Farben did in chemicals, and, in the United States at least, could play one chemical company against another. Armed with information and assured of co-operation which only a government can give, it could proceed intelligently and aggressively in contrast with a corporation limited to its own resources and confronted with antitrust legislation. Their objectives and therefore their motives were in partial contrast with each other. One pursued profit along with conquest by trade and force, while the other sought only financial gain; therefore, agreements by them to restrain production and sale—of magnesium, for example—was both market control and preparation for war in one case but only an incident to business profit in the other.

Evasion of Law. Another objective, a prerequisite to continued operation, is the appearance of legality in carrying on questionable activities. Antitrust legislation, tax laws, and especially the confiscation of enemy patents and other property in time of war cause cartel members, like Farben, to camouflage ownership of subsidiaries and holding companies and to employ understandings rather than written agreements. Manipulated sales through various corporations enabled Farben to get back most of the patents of United States subsidiaries seized by the United States during World War I.

According to Wendell Berge, in 1945, "We know that efforts were made to avoid seizure of German holdings before the outbreak of this war, and that camouflaged transfers of stock ownership, assignments of patents, and concealment of interest through the use of dummy firms were resorted to. Many such instances have been uncovered."[15]

[14] 77 Cong., 2 sess., *Senate Doc. 157,* 22.

General Aniline was organized in the United States and its principal officers were Germans who had become naturalized American citizens. In 1941, the United States seized over 90 per cent of the stock in General Aniline but later however the so-called Interhandel Corporation (formerly I. G. Chemie) of Switzerland filed suit against the government claiming ownership of most of the seized stock, thus preventing its sale until the point was adjudicated. This government contended that Interhandel is a dummy corporation controlled by Farben and utilized to conceal German ownership or control and thus prevent seizure of Farben properties. It was also alleged that Farben utilized General Aniline to conceal its ownership of important patents in the United States by assigning them to this "Swiss-owned" American firm before the war. In February, 1953, the United States District Court for the District of Columbia dismissed the complaint brought in *I. G. Chemie* v. *Brownell* because of I. G. Chemie's failure to comply with the court's order to produce certain records. The opinion of the district court in this case is reported at 111 F. Supp. 435. I. G. Chemie appealed the decision, but in June, 1955, the United States Court of Appeals for the District of Columbia unanimously affirmed the ruling of the lower court.

A Consistent Interpretation

The judicial interpretation of patent cartels, beginning as it did only a few years ago, has been consistent. Practically all of it comes within those years in which the Supreme Court shifted its emphasis from patent rights to antitrust laws in case of a conflict between them. In fact, decisions involving patent cartels are generally parts of opinions that condemn patent pools and consolidations and license agreements within this country. An examination of these cartels, their policies and effects, reveals their relationship to other forms of patent maneuvers and the place they have occupied in patent litigation.

Beryllium. Siemans and Halske Company in Germany originated beryllium. An addition of as little as 2 per cent of it to copper makes an alloy of unusual strength and hardness and therefore of particular importance in war industries. In 1934, Siemans and Halske, according to a government report, entered into a cross-licensing agreement

[15] EGRW Hearings, Part 4, 563.

with Beryllium Corporation which provided for the division of markets and other restrictions.[16]

Hydrogenated Gasoline and Synthetic Rubber. After World War I research revealed a close relation between the oil and chemical industries in that petroleum could be converted into rubber, explosives, fertilizers, alcohols, solvents, and plastics, and gasoline could be made through chemical processes. Both Standard Oil of New Jersey and Farben of Germany developed the hydrogenation process—making oil from coal or refining petroleum by means of new catalysts in the presence of hydrogen at high temperatures. According to the Bone Hearings and a government indictment, each company threatened the leading position of the other, Standard with its chemical derivatives from petroleum and Farben with its hydrogenated gasoline. Consequently, in 1929, they entered into a division-of-fields agreement by which Standard gave up its future chemical development not closely related to its business as of that year, and Farben withdrew from the petroleum industry outside of Germany. In addition, Standard gave to Farben approximately $30,000,000 worth of its shares of stock. The only chemical industries which Standard retained were alcohols and solvents, which it already had in 1929. Any withdrawal of each in favor of the other was to be accomplished by granting exclusive patent licenses and know-how, the chemical going to Farben and the petroleum to Standard. This division of fields, which pertained primarily to products made from petroleum, coal, and natural gas, applied, to repeat, only outside Germany; within that country Farben gave up no patent rights in hydrogenated gasoline although it acquired additional patent rights in butyl rubber and other chemicals. This agreement of 1929 was no mere pooling or licensing of patents as most patent cartels are but instead a consolidation of patent control on a world-wide basis, each one giving its patent rights in one field for the exclusive patent rights in another.

The division of fields was easier stated than administered, and these two companies had the problem of maintaining the interest of each in the other's technological progress. Many chemical products, both old and new, were related to the oil industry in different degrees —a situation which required supplementary arrangements. The Jasco

[16] TNEC Report, 8–9.

agreement, for example, provided that new chemical processes which use petroleum, coal, and natural gas as a starting material would be shared by both Standard and Farben through a jointly owned company, 62.5 per cent of the royalties going to the originator of the process and 37.5 per cent to the other. Even under the 1929 agreement, though Standard had full control of the oil and coal hydrogenation processes outside of Germany, Farben was allowed a 20 per cent financial interest to maintain its incentive in research to improve them.

Standard organized the Hydro Patents Company to exploit its complete control of the hydrogenation patents in the United States. The major oil companies were induced to become Hydro's licensees, to transfer to it their present and future patents on this process, and to permit it to inspect their plants and learn of new developments in refining. Their restricted use of this process was apportioned according to their daily crude output, and Hydro's patents were administered to prevent any use of coal as a raw material in making petroleum products. The obvious purpose of Standard of New Jersey was to become dominant again in the oil industry as another Standard company was prior to a dissolution decree in 1911.

In 1930, Standard and Farben orally agreed that all processes for producing synthetic rubber would be exploited jointly under the latter's control. Farben developed the buna type of synthetic rubber and began the large-scale production of it in 1934. Rubber companies were encouraged to believe that their requests for licenses would be granted. The apparent purpose was to discourage their independent research and to find out their progress toward synthetic rubber. In 1938, Standard developed a new process for making synthetic rubber, called butyl. It revealed this method to Farben, but failed to receive in return corresponding know-how about the buna process. On September 25, 1939, right after the outbreak of World War II, Farben assigned to Standard its buna patent rights in the United States and the British and French empires but still withheld the know-how. Under orders from the Hitler government, Farben refused to permit others to exploit synthetic rubber. In a cable dated October 16, 1939, it declared, "Referring to your question with respect to technical information about buna we have to inform you that under present conditions we will not be able to give such information."[17]

In November, 1939, Standard notified rubber companies that it had acquired control of buna patents and would license them at a certain rate per pound provided they gave it control of their patented improvements for manufacturing rubber and produced such rubber for their own manufacturing needs but not for sale in bulk. These restrictions were designed to permit Standard and Farben to continue their dominance in synthetic rubber. Standard's legal counsel commented August 21, 1942, on the proposed agreement: "This is not a cross-licensing agreement, but one in which patents are piled on patents in the hands of one centralizing company."[18] Firestone and U. S. Rubber accepted licenses. In a memorandum, a Standard official stated that Goodrich, which had done considerable research of its own, had applied for a license. "They have definitely stated that they do not need any technique from Germany but require only a bare patent license—and that if refused a license, they are going to proceed anyway with alternative processes which they think will be non-infringing."[19] The government interrupted the scheme by asking Standard to enter into a pool with other rubber companies to license buna patents and thus promote the defense program; this was done in December, 1941. However, about three years were required to make buna rubber in the United States in substantial amounts and of proper quality, although such production had already been achieved in Germany.

In 1938, officials of Farben had become also officials of the German government, and Farben's relations with Standard were subordinated accordingly to German plans for economic and military conquest; in short, they would not reveal strategic processes and products to non-Germans. However, Standard remained a business corporation with no interest except its own to serve. As a licensor and as a member of a United States pool covering the polymerization process, it, unknown to its associates, kept Farben informed about hydrogenation and kindred developments in this country. Standard's desire to strengthen its position in hydrogenated gasoline largely explains its

17 Bone Hearings, Part 6, 2678; Parts 6–9 are sources of information about Standard and Farben.

18 *Ibid.*, 2962.

19 *Ibid.*, 2682.

complete fulfilment of contractual promises to Farben in spite of Farben's failure to reciprocate.

In 1942, a government indictment charged Standard and Farben with agreements to use their combined present and future patents to prevent others from manufacturing and selling better and cheaper oil and chemical products. A consent decree of the same date required Standard to sever completely its relations with Farben, and provided for free public licensing during the war of all their patents covering the synthetic production of gasoline and rubber and for furnishing know-how to licensees under these patents. A supplemental judgment of 1943 provided for compulsory licensing, at a reasonable royalty, of the catalytic refining patents owned or applied for by Standard on March 15, 1942. These patents were especially important in the production of high-octane aviation gasoline.[20]

Aluminum. Alcoa participated in cartels in which foreign manufacturers agreed not to bring aluminum to the United States or to import it only under price and other restrictions. A circuit court of appeals found that Aluminium Limited, a Canadian subsidiary, had made agreements with European concerns which restricted aluminum imports into the United States.[21]

Titanium. Titanium compounds are suitable for the manufacture of white paint and valuable in the production of rubber, glass, and paper. About 1914, three groups of chemists, independent of each other but with a common background of science and technical arts, invented three different processes for making them. The American group assigned their patents to Titanium Pigment, the Norwegian to Titan, and the French to Terres Rares. In 1920, Titanium Pigment, affiliated with National Lead Company, entered into an agreement with Titan of Norway that became the basis for an international cartel in titanium pigments. The American and Norwegian companies exchanged patent rights through exclusive licenses on present and future patents, each accepting the validity of the other's patents. They divided the earth into sales territories for their exploitation;

[20] *United States* v. *Standard Oil Co.* (*N.J.*), CCH Trade Regulation Reports, Supp. 1941–43 ¶ 52,768 and ¶ 52,927 (D.C.N.J. Civil No. 2091; Complaint and Consent Decree, March 25, 1942; Supp. Judgment, 1943). A recent account of these hydrogenation patents is found in *United States* v. *Standard Oil Co.* (*N.J.*), *et al.* (D.C. of D.C. Civil No. 86–27; Complaint, 1953).

[21] *United States* v. *Aluminum Co. of America,* 148 F. 2d 416 (2nd Cir. 1945).

Titanium had the North American continent, and Titan the remainder of the globe except South America—a common market for both of them. Each imparted technological information about titanium to the other and permitted plant inspection. In 1927, National Lead acquired 87 per cent of Titan's stock and, in 1929, a subsidiary through which National acted in the foreign field, acquired the rights and assumed the liabilities of Titan under the 1920 contracts. Farben, ICI, and other important chemical companies throughout the world joined in carrying out the scheme to restrain international commerce. In 1933, Du Pont, after having acquired patents on an alternative method in this field, joined the conspiracy, although its rights and obligations were somewhat different from those of other members of the combination. Beginning in that year Du Pont and National Lead were the only producers of titanium pigments in the United States. The avowed objectives of the entire arrangement were the elimination of competition and the exchange of technology. From 1920 to 1940, the production and sale of titanium pigments forged ahead of lithopane and white lead—competing products—and prices were repeatedly reduced.[22]

In 1944, the government sued National Lead, Titan, and Du Pont for violation of the antitrust laws. Foreign coconspirators, seventeen in number, were not cited as defendants because of lack of jurisdiction. A district court, in 1945, held that these three companies had violated the antitrust laws, and the Supreme Court, in 1947, upheld the decision.[23] This judicial conclusion was reached from three approaches—the agreement of 1920, the qualified participation of Du Pont in the combination after 1933, and the contract between National Lead and Du Pont apart from their relation to ICI and other foreign producers. Both courts—district and Supreme—emphasized the great power which these two companies acquired, their domination over titanium pigment in the United States by cross-licensing hundreds of related patents, and the absence of litigation to test the validity of any of them. "These patents, through the agreements in which they are enmeshed and the manner in which they have been used, have, in fact, been forged into instruments of domination of an entire industry. The net effect is that a business, originally founded

22 *United States* v. *National Lead Co.*, 63 F. Supp. 513, 517–521 (S.D.N.Y. 1945).

23 *Ibid.*, and 332 U.S. 319 (1947).

upon patents which have long since expired, is today less accessible to free enterprise than when it was first launched."[24] The decree of the district court included cancellation of restrictive agreements with either domestic or foreign corporations and uniform licenses at reasonable royalties to any applicant under their existing patent rights, including twenty-one of Farben's exclusively licensed to National Lead and vested in the alien property custodian.[25] In 1920, the foundation was laid for restraint of interstate commerce and foreign trade; in 1944, the government brought action against the conspirators in this country; and in 1947, the Supreme Court upheld the decree of the lower court.

Tungsten Carbide. Tungsten carbide, one of several hard metal compositions, is superior to steel and other materials as an edge for hard cutting tools. According to a district court, General Electric, its two subsidiaries named Carboloy and International, and Krupp of Germany, "owned and acquired patents which were so used as substantially to dominate the entire interstate and foreign commerce of the United States in the field of hard metal compositions. This situation continued from the inception of the industry in this country in 1928, until 1940, when some of the patents were declared invalid in an infringement suit in Michigan."[26] In the 1928 contract, "basically, Krupp assigned to GE all its present and future United States patent rights in the field of hard metal compositions."[27] In addition, "Krupp itself was permitted to import into the United States under any of the GE-held patents"; GE agreed to fix the prices of its licensees and Krupp promised to conform to these prices.[28] GE was to receive one-third and Krupp two-thirds of the royalties of five dollars per pound which they, as well as licensees, paid.

The purpose of the plan is revealed in a report of American attorneys to Krupp concerning their conversation with GE's lawyer. "Each company has something to bargain with and it looks as though there is a real inducement for combining forces since neither company can dominate the situation alone and there is a fair chance that by pooling the patents competition could be discouraged."[29] However, GE per-

[24] *Ibid.*, 532.
[25] *Ibid.*, 532–35.
[26] *United States* v. *General Electric Co.*, 80 F. Supp. 989, 993 (S.D.N.Y. 1948).
[27] *Ibid.*, 995.
[28] *Ibid.*, 995.
[29] *Ibid.*, 995.

haps could have used its patents without infringing on Krupp's; and if it had contested the Krupp patents, which were finally held invalid, its monopoly might have been more complete than under the cartel. In 1936, the parties amended the contract of 1928 and as a result Krupp agreed not to import into the United States and GE not to export from this country except to Canada. Carboloy, GE's subsidiary, purchased Prosser and Sons in 1936 and Simons Union Wire Die in 1937 in order (1) "to effectuate the 1936 agreement removing Krupp from the U. S.-Canadian market, and (2) prevent the competition which Simons was able to offer aside from his position as a Krupp outlet."[30] Carboloy and its five manufacturing licensees headed the commercial scheme for exploiting the tungsten carbide patents in the United States. Two of these licensees were Ludlum Steel and Firth-Sterling to which Krupp had promised patent rights, the other three "competitors operating under patents which Carboloy always claimed were dominated by its own patent rights."[31] These three companies became licensees after being threatened with infringement suits, none of which was brought to trial from 1927 to 1940. In the cross-licensing of patents, the agreements provided that Carboloy would fix selling prices. Carboloy and the licensees sold their output either as blanks or as completed tools, dies, and so forth. They sold finished products and parts directly to users and also to so-called agents. These agents either resold the parts or products or made finished products out of the parts and sold them. At first the price of tungsten carbide, much higher in the United States than in Germany, discouraged the use of it; therefore when war came, in 1941, industry was using tungsten carbide tools much less here than over there. After the indictment of Carboloy, and after Ford was able to secure tungsten carbide elsewhere, the price dropped from $205 to $50 per pound.[32] This is perhaps the most striking example of a decrease in price following antitrust enforcement.

In response to the government charge that defendants "pooled independent competing patents," the court said, "The monopolistic price-fixing scheme in which they were engaged is illegal regardless of the relation between the patents they accumulated."[33] The court

[30] *Ibid.*, 1009–1010.
[31] *Ibid.*, 1000.
[32] Bone Hearings, Part I, 52.
[33] *United States* v. *General Electric Co.*, 80 F. Supp. 989, 1012 (S.D.N.Y. 1948).

condemned the entire arrangement between GE and Krupp, and between GE and its licensees and agents, for violation of antitrust legislation. This case, like the one involving lamps, was postponed in the name of war emergency although GE continued an infringement suit against Sylvania. The decision of a district court in 1948, holding the defendants guilty and imposing fines, came seven years after the indictment.

Incandescent Lamps. Since 1919, according to a report of the Tariff Commission published in 1939, International General Electric, GE's subsidiary for handling foreign business, "has entered into numerous agreements with foreign companies which provide for the exchange of patent licenses and manufacturing information, and for the establishment of territorial limits to competition between the parties to the agreements. Under the terms of the agreements now in effect, the companies apportion among themselves the principal world markets. . . . In return for the grants which it makes to the foreign companies, the General Electric Company receives from each of the companies an exclusive license to make and sell lamps in the United States under all patents owned or controlled by these companies."[34]

In 1941, the Department of Justice filed a complaint, and, in 1949, a district court condemned the agreements of 1925 (Phoebus) and of 1941 for the stabilization of the lamp industry throughout the world.[35] IGE was the manipulator which brought this cartel into being. It "negotiated the individual foreign license agreements each containing territorial restraints keyed to the Phoebus agreement and all requiring subsidiaries and licensees of the parties to respect the territorial limitations."[36] The government charge also said that "the motivating factor governing all IGE foreign contracts from its very inception was the intent to maintain the dominant position of General Electric in the United States by protecting it from foreign competitors."[37] In a letter of 1926 from the vice-president to the president of GE the statement is made: "You are familiar with the work that has been done and the plans that have been developing in foreign countries during the past twenty-four years, all for the purpose of protecting our domestic lamp business against foreign encroachments

[34] *Incandescent Electric Lamps,* Tariff Commission *Report No. 133* (1937), 59.

[35] *United States* v. *General Electric Co.,* 82 F. Supp. 753 (D.N.J. 1949).

[36] *Ibid.,* 843.

[37] *Ibid.,* 843.

when our important patents expire."[38] GE "activities in the United States were geared to the Phoebus agreement and were controlled by virtue of it."[39] GE utilized the export provisions of its licenses to Westinghouse and other domestic manufacturers as a means of preventing lamp sales in foreign fields. Moreover, it refused to grant domestic manufacturers licenses under its foreign patents. "The interest of the public in the world-wide divisions of territory set up and the far flung effects upon competition and trade encompassed in them is very great." The evidence showed that "a lively competition might have developed but for the territorial restrictions placed upon the General Electric domestic licensees."[40] GE "paced its industrial achievements with efforts to insulate itself from competition. It developed a tremendous patent framework and sought to stretch the monopoly acquired by patents far beyond the intendment of those grants. It constructed a great network of agreements and licenses, national and international in scope, which had the effect of locking the door of the United States to any challenge to its supremacy in the incandescent electric lamp industry arising from business enterprise indigenous to this country or put forth by foreign manufacturers."[41] Consent decrees of Westinghouse and Corning, previously described, were in keeping with this decision. A final judgment of a district court in 1953, already presented in Chapter Four, required the dedication to the public of certain United States patents on incandescent lamps foreign owned as well as kindred patents owned by GE and other defendants.[42]

Coated Abrasives. Minnesota Mining and Manufacturing, Carborundum, and other manufacturers of coated abrasives organized Durex Corporation under the Webb-Pomerine Act and licensed it under their foreign patents. Durex controlled English, German, and other foreign companies which made and sold coated abrasives, and the domestic producers agreed in effect not to export to those countries—an arrangement, according to a district court, in conflict with the Sherman Act. "Since the patent pool was avowedly ancillary to a plan which this Court finds to be a forbidden restraint of trade, the

[38] *Ibid.*, 811.
[39] *Ibid.*, 843.
[40] *Ibid.*, 847.
[41] *Ibid.*, 905.
[42] *United States* v. *General Electric Co.*, 115 F. Supp. 835 (D.N.J. 1953).

main and subsidiary patent agreements are invalid *in toto*."[43] A part of the decree required the defendants to issue unrestricted licenses under royalty arrangements.

Chemicals. Du Pont, Imperial Chemical Industries, and Farben, or their predecessors, divided world markets on chemical products, sporting arms, and ammunition, first in 1887 by simple agreement, and later, in 1907 and 1913, on the basis of an exchange of "present and future patents, processes and inventions relating to military and commercial explosives."[44] The agreement of 1920 included only Du Pont and ICI; an official of ICI called it "a camouflage to cover all relationships between the two companies."[45] The 1929 and 1939 agreements accomplished the same general purpose. Remington Arms Company became a part of this arrangement in 1933 when Du Pont acquired most of its stock. Du Pont and ICI, both with footholds in Argentina and Brazil, combined in 1933 to form Duperial-Argentina and Duperial-Brazil, "to the end that the same foreign market may be developed for their mutual benefit and profits divided on an agreed basis."[46]

A district court found that the various patents and processes "agreements, irrespective of their per se legality, were instruments designed and intended to accomplish the world-wide allocation of markets; their object was to achieve an unlawful purpose—an illegal restraint of trade prohibited by section one of the Sherman Act."[47] The court also found "that the jointly-owned companies were means designed and used by Du Pont and ICI to avoid and prevent competition between themselves and with others in the nonexclusive territories."[48] In dismissing a contention of the defendant the court said, "There is no intimation in any decision that elimination of competition is to be given a more favorable judicial consideration when achieved by the route of territorial division rather than by way of price fixing, or that proof of industry domination is required in one

[43] *United States* v. *Minnesota Mining and Mfg. Co.*, 92 F. Supp. 947, 964 (D. Mass. 1950); *ibid.*, 96 F. Supp. 356 (D. Mass. 1951).

[44] *United States* v. *Imperial Chemical Industries, Ltd.*, 100 F. Supp. 504, 516 (S.D. N.Y. 1951).

[45] *Ibid.*, 516.

[46] *Ibid.*, 557.

[47] *Ibid.*, 592.

[48] *Ibid.*, 592.

case though not required in the other."[49] The court also said, "Similarly do we deem irrelevant any inquiry into whether the arrangements between the parties actually injured the public interest, or whether the public benefited thereby."[50] In support of these statements the court quoted from the condemnation of a roller bearing cartel by the Supreme Court.[51] The English courts refused to observe the terms of the decree—a fact which illustrates the lack of complete jurisdiction of any one country and therefore the difficulty in enforcing a judicial decision against a cartel.

Other Products. According to complaints and consent decrees, certain products, already discussed in the chapters on pools, and consolidations and license agreements, were also subject to cartel control. They include magnesium, tempered glass, lecithin, fluid-filled cables, duplicating machines and stencils, gas-refrigeration equipment, screws and drivers, glass fibers, and alloys and dentures. The same source of information points to additional examples.

a. *Military Optical Instruments.* Bausch and Lomb in this country and Carl Zeiss of Germany did not compete in the sale of military optical instruments, one having the United States to itself and the other the remainder of the world. A consent decree of 1940, with the same date as the complaint, enjoined the agreement.[52]

b. *Glass Bulbs.* In 1940, according to an indictment of that year, General Electric and Corning, sole manufacturers of electric glass bulbs in the United States, paid Dutch companies between $15,000 and $25,000 per year to insure freedom from competition in North America. In 1941, they pleaded *nolo contendere* and fines were imposed upon them.[53]

c. *Nitrates.* According to a government complaint of 1941, Synthetic Nitrogen Products Corporation, organized to handle all nitrogen-bearing fertilizer made by Farben and sold for importation into the United States, and other producers of fertilizer nitrates conspired with Farben to restrict the use of certain patents and licenses to the

49 *Ibid.,* 593.

50 *Ibid.,* 593.

51 See *Timken Roller Bearing Co.* v. *United States,* 341 U.S. 593 (1951).

52 *United States* v. *Bausch and Lomb Optical Co.,* CCH Trade Regulation Reports, Supp. 8th ed., Vol. 3, ¶ 25,487 (S.D.N.Y. Civil No. 9–404; Complaint and Consent Decree, July 9, 1940).

53 *United States* v. *Corning Glass Works* (S.D.N.Y. Cr. No. 108–164; Indictment, 1940; Fines, 1941).

manufacture of ammonia for agricultural fertilizers only. A consent decree of the same date enjoined these practices and forbade any restrictions on the use of certain United States patents or the paying of any royalties for the use of them.[54]

d. *Pharmaceuticals and Drugs.* Between 1920 and 1926, Bayer, a subsidiary of Sterling Products, entered into agreements which permitted Farben of Germany to share in its profits from export business. Also, Winthrop Chemicals, another subsidiary of Sterling, acquired the exclusive right to use Farben patents, trade-marks, and secret methods of manufacturing new pharmaceuticals and refrained from exporting them to certain foreign markets. In 1926, Winthrop issued one-half of its stock to Farben and cancelled its previous obligation to divide its profits. Alba Pharmaceutical also participated in these cartel restrictions. A consent decree in 1941, in response to a complaint of the same date, terminated these agreements, and Sterling and its subsidiaries were ordered not to make any agreement to stay out of foreign markets.[55]

Certain hormones, manufactured synthetically within recent years and used to counteract shock from wounds and burns, have been produced and sold according to cartel agreements. According to a government complaint of 1941, four domestic companies conspired with two European corporations to prevent others from manufacturing hormones in the United States or importing them, to allocate among themselves various countries as markets, and to maintain arbitrary prices by withholding patent rights from competitors and by having restrictions in license agreements. A consent decree of the same date forbade the carrying out of these illegal practices.[56]

Another government complaint, in 1943, charged that Merck and Company entered into an agreement with E. Merck Chemical Works of Germany for a division of world markets and an exchange of patents and technological information for pharmaceutical products. A consent decree in 1945 enjoined Merck from giving Merck of

[54] *United States* v. *Synthetic Nitrogen Products Corp.*, CCH Trade Regulation Reports, Supp. 1941–43 ¶ 52,700 (S.D.N.Y. Civil No. 15-365; Complaint and Consent Decree, Sept. 5, 1941).

[55] *United States* v. *Alba Pharmaceutical Co.*, CCH Trade Regulation Reports, Supp. 1941–43 ¶ 52,650 (S.D.N.Y. Civil No. 15-363; Complaint and Consent Decree, Sept. 5, 1941).

[56] *United States* v. *Schering Corp.*, CCH Trade Regulation Reports, Supp. 1941–43, ¶ 52,729 (D.C.N.J. Civil No. 1919; Complaint and Consent Decree, Dec. 17, 1941).

Germany any control over its affairs and from exchanging patents with it upon condition of refraining from competition. It further required Merck to license about sixty patents royalty-free and to surrender for license by the alien property custodian its rights to fifty patents formerly owned by the German company.[57]

e. *Fuel Injection Equipment, Aircraft Magnetos, and Automotive Electrical Equipment.* According to a complaint of 1942, American Bosch Corporation and certain foreign companies exchanged patent licenses and divided sales territories relating to these products. A consent decree of the same date cancelled various agreements; required unrestricted licenses for certain patented products without conditions, except the payment of a reasonable royalty after the war; and enjoined American Bosch from transferring patent rights except with the court's approval.[58]

f. *Aircraft Instruments and Parts.* Between 1935 and 1940, according to a complaint of 1942, Bendix Aviation, a manufacturer and distributor of aircraft instruments and parts, entered into agreements with foreign companies pertaining to markets, prices, patents, and know-how. License agreements with British companies covered such things as starters, magnetos, generators, pumps, and spark plugs. The Italian agreement pertained to aerophotographic apparatus and the Japanese to "rate of change in altitude" indicators. German agreements covered instruments, de-icers for removing and preventing ice, nozzles for injection of fuel, and supercharger governors. Royalty terms varied from one agreement to another according to the relative importance of patents and other factors. Bendix apparently received royalties in more instances than it paid them, which indicates the comparative importance of its patents and know-how. In 1946, a consent decree enjoined these cartel agreements, forbade any suit for patent infringement prior to this date, transferred title in 136 Bendix patents to the alien property custodian for granting royalty-free licenses under them, and required Bendix to issue licenses at reasonable royalties under 144 additional patents.[59]

[57] *United States* v. *Merck and Co.*, CCH 1944–45 Trade Cases ¶ 57,416 (D.C.N.J. Civil No. 3159; Complaint, 1943; Consent Decree, 1945).

[58] *United States* v. *American Bosch Corp.*, CCH 1948–49 Trade Cases ¶ 62,284 (S.D.N.Y. Civil No. 20–164; Complaint and Consent Decree, Dec. 29, 1942).

[59] *United States* v. *Bendix Aviation Corp.*, CCH 1946–47 Trade Cases ¶ 57,444 (D.C.N.J. Civil No. 2531; Complaint, 1942; Consent Decree, 1946).

g. *Matches*. Swedish Match, which played a dominant role for many years, had seventy-five subsidiaries and several partly owned companies in over thirty different countries. According to a government complaint of 1944, Swedish Match was the world's largest producer of matches and of lumber, chemicals, and machinery for their fabrication and, prior to 1939, was the principal exporter. Holding companies closely affiliated with it until their bankruptcy in 1932 were International Match and Kreuger and Toll; Ivar Kreuger was the leader from 1917 to 1932. Diamond Match, first organized in 1880 as a consolidation of thirty-five match companies, acquired other companies; British Match, partly owned by the Swedish and Diamond companies, acquired the stock of Bryant and May—virtually the sole producer of matches in Great Britain. Eddy Match of Canada was owned 95 per cent by Diamond and Bryant and May.

These consolidations and stockholdings led to control of markets and patents. Diamond, for example, refrained from exports and became the sales agent for the limited number of matches imported from Sweden and other countries. According to an agreement of 1922, Diamond purchased chlorate of potash for its own use and for resale from the predecessor of Farben and did not manufacture it on a commercial scale; Farben sold to no one else here. Diamond scrapped its plants for making this chemical, which is essential to ammunition as well as matches. This agreement may well have been another phase of the scheme for German world conquest, first in peace and then in war.

Diamond also acquired control of match machinery in the United States partly through the ownership of patents. The Swedish and Bryant and May companies exercised similar control over match machinery in other countries. All these companies, of course, interchanged patent rights and technological information.

Beginning in 1932, several basic and improvement patents on a so-called everlasting match were issued in fifty-one countries to the inventors of it. In the same year, Kreuger secured control of the patents on this match for the Swedish company. Later Bryant and May paid substantial sums for an interest. However, Diamond decided to leave it alone—"It is to be hoped that if the item is not put out and pushed by a strong manufacturer, no one else will take it up even after the patents expire."

A consent decree of 1946 provided for royalty-free licensing of Diamond patents and applications and of the Swedish patents on the everlasting match. For five years—until 1951—Diamond had to furnish know-how relating to match machinery to anyone at a nominal charge. Also it could not sell Swedish matches nor acquire any domestic manufacturer unless it could be shown that competition would not be unreasonably restricted. Other provisions sought to terminate the control of corporations through stock ownership.[60] One may ask, though, why the defendant—in this case Diamond Match—accepted royalty-free licensing in view of the disapproval of such a measure by the Supreme Court in the Hartford-Empire case?

h. *Electric Precipitators*. Particles suspended in gases may be precipitated by passing the gases through an electrical field. The invention for doing this, the basic patent on which was granted to Cottrell in 1907, recovers practically all the cadmium produced in this country and it is now used in connection with the manufacture of high-octane gasoline, synthetic rubber, aluminum, and steel. According to a government complaint of 1945, International Precipitation, a patent holding company, granted exclusive licenses to Western Precipitation and Research Corporation to manufacture and install electrical precipitators and the component parts for them. The remainder of the world was divided by International among five foreign companies; for example, the British Empire was assigned to Lodge-Cottrell, Ltd. of Birmingham. The members of the cartel sought to control all new developments in the art by securing the patent rights to all improvements from their employees and their licensees and by purchasing them from outsiders. They co-operated in research, cross-licensed each other, and agreed not to obtain patents in each other's territory. Although the last of the three basic patents expired in 1937, the cartel members continued by agreements to control all phases of the electrical precipitation business, unpatented as well as patented. From 1930 to 1945, Western Precipitation and Research had approximately 90 per cent of the business in the United States. Their suit for infringement against an engineering firm which developed an electrical precipitator for cleaning blast furnaces was settled by both sides cross-licensing each other. A consent decree in 1946 prohibited the

60 *United States* v. *Diamond Match Co.*, CCH 1946–47 Trade Cases ¶ 57,456 (S.D. N.Y. Civil No. 25–397; Complaint, 1944; Consent Decree, 1946).

allocation of markets and restrictive license agreements, enjoined the enforcement of United States and foreign patents in this field, and required all present know-how be made available through deposit of manuals in university libraries.[61] In this case, as in others, it is not clear why the defendant accepted such a patent provision.

i. *Storage Batteries.* The Electric Storage Battery Company of Philadelphia had three subsidiaries—two, Willard of Cleveland, Ohio, and Exide of Canada, wholly owned; one, Chloride of England, partly owned. According to government complaint in 1945, Electric Storage, its subsidiaries, and a German company, in a series of agreements beginning in 1891, divided certain world markets and fixed prices in nonexclusive territories. These companies exchanged all pertinent patents and unpatented inventions and technology with each other for exclusive use in their respective territories. Their basic patents had expired in 1915. Electric Storage and Chloride used the same trademarks and in nonexclusive territory divided the supervision of areas and all profits. One alleged effect of this cartel was to deprive the United States of cadmium batteries either through manufacture or imports. A consent decree in 1947 enjoined Electric Storage and Willard from maintaining these restrictive agreements, and ordered the former to grant nonexclusive licenses under several of its patents to any applicant upon reasonable terms and also to transfer its voting rights in Chloride to a trustee.[62]

j. *Plastics.* Acrylic products, an important kind of plastics, are derived by chemical synthesis from coal, petroleum, air, and water. In cast sheets they are used for cockpit enclosures; in powder form, for tableware, furniture, dentures, and ornaments; and in emulsion, for adhesives, laminating layer in shatterproof glass, and an intermediate in the manufacture of synthetic rubber. Rohm and Haas of Philadelphia, beginning in 1934, entered into restrictive agreements with Farben, Rohm and Haas of Germany, and ICI of Great Britain. Rohm and Haas and Farben, for example, licensed each other under their respective patents and exchanged technical information. In addition, the two German companies and ICI did not export acrylic prod-

[61] *United States* v. *Western Precipitation Corp.,* CCH 1946–47 Trade Cases ¶ 57,458 (S.D. Calif. Civil No. 4677; Complaint, 1945; Consent Decree, 1946).

[62] *United States* v. *Electric Storage Battery Co. and Willard Storage Battery Co.,* CCH 1946–47 Trade Cases ¶ 57,645 (S.D.N.Y. Civil No. 31–225; Complaint, 1945; Consent Judgment, 1947).

ucts to the United States or fabricate them here. Rohm and Haas of the United States refrained from exporting certain acrylic products, and from making, using, or selling them in or outside the United States in specified commercial fields reserved to Farben. In return Farben did not make, use, or sell such products anywhere for the preparation of shatterproof glass and adhesives. There was a division of markets elsewhere between British and German companies.

A consent judgment of 1948, in response to a government complaint of the same date against Rohm and Haas, cancelled these cartel agreements, and required the licensing royalty-free of seventy-four Rohm and Haas and ninety foreign-owned patents on making acrylic products and required the furnishing of technical information until 1951.[63] In this case, the office of the alien property custodian, which owned Rohm and Haas and was in the process of disposing of it, played an active part.

k. *Television.* According to a government complaint of 1945, Scophony of America, in 1942, acquired control in the United States of the Scophony Ltd. invention of television which was owned by two companies—General Precision Equipment and Television Productions, both affiliated with motion-picture companies—a fact which may explain their failure to develop this invention. A consent decree of 1949 required these two companies to get rid of their Scophony stock and to terminate their exclusive licenses under Scophony patents and processes, and enjoined the division of the world into noncompetitive areas. A similar judgment was entered against Scophony Ltd.[64]

l. *Photoprinting Materials.* In 1941, General Aniline and Film Company and one of its officers were charged with restraining trade in photoprinting materials by means of exclusive sales agreements, patent licensing restrictions, and allocation of world markets; in 1950, they pleaded *nolo contendere,* and moderate fines were imposed.[65]

[63] *United States* v. *Rohm and Haas Co.,* CCH 1948–49 Trade Cases ¶ 62,334 (E.D. Pa. Civil No. 9068; Complaint and Consent Judgment, Nov. 18, 1948).

[64] *United States* v. *Scophony Corp.,* CCH 1948–49 Trade Cases ¶ 62,238, ¶ 62,356 and ¶ 62,463 (S.D.N.Y. Civil No. 34–184; Complaint, 1945; Consent Decree, 1948 and 1949). The Supreme Court held that Scophony Ltd. was within the jurisdiction of a lower court (*United States* v. *Scophony Corp. of America,* 333 U.S. 795 (1948); reversing 69 F. Supp. 666 (S.D.N.Y. 1946).

[65] *United States* v. *Dietrich A. Schmitz,* CCH 1944–45 Trade Cases ¶ 57,305 (S.D. N.Y. Cr. No. 111–137; Indictment, 1941; Imposition of Fines, 1950).

m. *Stud-Welding Equipment.* In 1949, the Department of Justice charged Morton Gregory Corporation of Ohio with licensing several foreign companies under its patents on stud-welding equipment. Also, each licensee, allocated a certain territory for its operations, refrained from making and selling any competitive device and used the same trade-mark. Morton Gregory reserved the United States market for itself. A consent judgment of 1951 enjoined the allocation of territories and restrictions on manufacture and sale.[66]

n. *Water-conditioning Apparatus and Materials.* According to a government complaint of 1945, the leading manufacturers and distributors of water-conditioning products, including the Permutit Company of New York and coconspirators of England, Germany, France, Belgium, and Spain, divided world markets, prevented all others from using their patent rights and technical information, and utilized the trade name "Permutit" to facilitate their scheme. A consent judgment of 1951 provided for the sublicensing of patents, ten of them royalty-free and one hundred twenty-five on a reasonable royalty basis, and for exports from and imports into the United States regardless of foreign or domestic patents. Permutit was also required to adopt a different trade-mark for foreign trade and to conduct its business in a way that would make possible the sale of its products abroad.[67]

o. *Rolling-Mill Machinery and Equipment.* Alleged facts in 1950 indicate that United Engineering and Foundry Company and thirteen coconspirators by allocating world markets eliminated competition in the manufacture and sale of rolling-mill machinery for working metals. A consent decree of 1952 provided, among other things, that United could not exchange patent rights or know-how with restrictions on imports into, or manufacture within, this country.[68]

p. *Electric Generators, Transformers, and Motors.* According to a government complaint of 1945, Westinghouse Electric and two "Siemens" companies of Germany divided markets for electric generators, transformers, motors, and appliances, and exchanged exclusively among themselves both rights and know-how under their present and

[66] *United States* v. *Morton Gregory Corp.,* CCH 1950–51 Trade Cases ¶ 62,750 (N.D. Ohio Civil No. 6279; Complaint, 1949; Consent Judgment, 1951).

[67] *United States* v. *Permutit Co.,* CCH 1950–51 Trade Cases ¶ 62,888 (S.D.N.Y. Civil No. 32–394; Complaint, 1945; Consent Judgment, 1951).

[68] *United States* v. *United Engineering and Foundry Co.,* CCH Current Trade Cases ¶ 67,378 (W.D. Pa. Civil No. 8926; Complaint, 1950; Consent Judgment, 1952).

future patents. A consent decree of 1953 required Westinghouse to dedicate approximately 250 United States patents acquired from the German companies and enjoined any future restrictive agreements between these companies.[69]

q. *Electrical Equipment.* General Electric and six coconspirators of Germany, France, England, Japan, Belgium, and Italy restrained foreign commerce in all types of electrical equipment except electric lamp and radio apparatus. They divided world markets and granted each other exclusive rights under all their present and future patents relating to such products. A final consent judgment of 1953 forbade these arrangements and required GE to grant licenses at reasonable royalties under approximately six hundred United States patents which it had acquired under the cartel agreements and also under its own patents on electrical equipment which are necessary to practice the inventions covered by these foreign patents.[70]

r. *Seamless-Hosiery Machine.* Scott and Williams, Inc. and an individual were charged with obtaining a monopoly in the seamless-hosiery machine industry through allocation of world markets, acquisition and suppression of patents, and misuse of patents in coercing competition. A consent judgment in 1954 enjoined such restraints as the division of territories, impediments to exports and imports, distributor relationships with coconspirators, and discriminatory and tying practices in the sale of such machinery.[71]

s. *Latex.* United States Rubber Company, Dunlop Rubber Company, and their jointly-owned rubber companies restricted to certain areas the manufacture and sale of latex and dipped goods, foam rubber, and thread; they and other foreign companies pooled their latex patents and processes and gave licenses, with restrictive conditions covering markets and prices, to other companies. A consent judgment required these two companies to grant, with reasonable nondiscriminatory royalties, nonexclusive licenses under their pooled foreign patents and related technical information, and to permit imports into

[69] *United States* v. *Westinghouse Electric Mfg. Co.,* CCH Current Trade Cases ¶ 67,501 (D.C.N.J. Civil No. 5152; Complaint, 1945; Consent Judgment, 1953).

[70] *United States* v. *General Electric Co.,* CCH 1952–53 Trade Cases ¶ 67,585 (D.C. N.J. Civil No. 45–75–C; Complaint, 1945; Final Consent Judgment, 1953).

[71] *United States* v. *Scott and Williams Inc.,* CCH 1953–54 Trade Cases ¶ 67,748 (S.D.N.Y. Civil No. 54–358; Complaint, 1949; Consent Decree, 1954).

this country by these licensees. Also it prohibited such objectionable practices as division of markets and price fixing.[72]

Complaints. Recent complaints of the Department of Justice deal with a variety of cartels. Bendix Aviation and its subsidiaries, Hydraulic Brake and Bendix-Westinghouse, and various foreign companies divided the world into noncompetitive areas for the manufacture and sale of braking systems and arbitrarily allocated these areas to themselves. As a result, the licensees of Hydraulic Brake patents were not permitted to export hydraulic braking systems for installation outside the United States.[73]

Manufacturers of prismatic glassware, and illuminating appliances containing it, allocated world markets and exchanged information about pertinent inventions, designs, and research.[74] Similar restraints involved alginate dental impression powder.[75]

Morgan Construction Company manufactured and sold wire-drawing machinery in the United States and assigned European rights to manufacture it to Malzine, a Belgium company, and the right to sell it to Belgameric of New York. Morgan Construction furnished Malzine with blue prints, drawings, and engineering assistance, and received by assignment any patent rights covering their improvements of it.[76] Beginning in 1937, three manufacturers of gear cutting and finishing machines and tools, by means of patent licensing and cross-licensing agreements, allocated territories, customers, and production among themselves, and fixed prices, terms, and conditions of sale.[77] Lastly, five manufacturers of railroad car-couplers divided world markets under agreement with English and Belgium companies.[78]

Conclusion

The German government sponsored Farben's program of cartel restrictions upon synthetic rubber, magnesium, and other strategic ma-

[72] *United States* v. *United States Rubber Co.,* CCH 1953–54 Trade Cases ¶ 67,771 (S.D.N.Y. Civil No. 50–564; Complaint, 1948; Consent Decree, 1954).

[73] *United States* v. *Bendix Aviation Corp.* (S.D.N.Y. Civil No. 44–284; Complaint, 1947).

[74] *United States* v. *Holophane Co.* (S.D. Ohio Civil No. 2659; Complaint, 1949).

[75] *United States* v. *L. D. Caulk Co.* (D. Del. Civil No. 1372; Complaint, 1951).

[76] *United States* v. *Morgan Construction Co.* (D. Mass. Civil No. 52–914; Complaint, 1952).

[77] *United States* v. *Michigan Tool Co.* (E.D. Mich. Civil No. 12605; Complaint, 1953).

[78] *United States* v. *National Malleable and Steel Castings Co.* (N.D. Ohio Civil No. 30281; Complaint, 1953).

terials. Also, other governments have been friendly or at least indifferent to certain cartels. In recognition of this situation and the need for meeting it with practical measures, the United States passed the Webb-Pomerene Act of 1918. It permits domestic companies to combine for export sales abroad, but forbids unfair competition in such trade that would result in adverse effects on the domestic market. Courts in the Alkali and Minnesota Mining cases held that an export association which is a cartel member is unlawful because restraint of trade is unfair competition.[79] In view of this interpretation one may conclude that the United States is the only nation which opposes international as well as national monopoly and restraint of trade in products. Antitrust legislation and the courts have condemned such monopoly and restraint although the executive branch has not always been aggressive in enforcing the law.

What accounts for this contrast between other nations and the United States? Older nations in the past have thrived on restrictions that permitted them to sell their finished products in their colonies. As these dependencies became self-governing or entirely independent, their opposition to such a program became firmly established. The United States in particular was born in a spirit of antagonism to economic restraints, and the federal government as well as various states still declare that monopoly is contrary to the spirit of a free people. An old civilization tends to accept the status quo and to become encrusted with custom—a favorable environment for restraints of all kinds. On the other hand, a relatively new country requires more initiative and creates individualism, both conducive to competition and even tolerance for its abuses. Old age, in contrast to youth, seeks safety rather than risk.

European countries still are lopsided in their economic life in that they are mainly producers of finished goods for foreign markets and purchasers of raw materials from other countries. In other words, a high proportion of their sales and purchases are abroad; therefore they take readily to a policy of selling high and buying low. They believe in selling manufactured goods under cartel agreements and in buying cotton and wheat and other raw materials under competitive conditions. Exceptions arise because of financial interests in the

[79] *United States* v. *United States Alkali Export Assn.*, 86 F. Supp. 59 (S.D.N.Y. 1949); and *United States* v. *Minnesota Mining and Mfg. Co.*, 96 F. Supp. 356 (D. Mass. 1951).

production and sale of such things as tin and natural rubber. In contrast, the United States produces both finished goods and the raw materials for them. Most of its sales and purchases are at home. A program of either selling high or buying low or both hurts a large proportion of people and therefore arouses opposition; witness, for example, the complaints about parity prices for farm products and the proposal to continue to buy them at such prices but to sell them on a competitive level and therefore please opposing groups of people.

Closely related to this point is the availability, extent, and effectiveness of information about economic matters and the public opinion that results therefrom. A policy of selling finished goods high and buying raw materials low within a country hurts those to whom such goods are sold and those from whom such materials are bought. What pressure can they exert for relief? In Europe people have little opportunity to learn and use unfavorable facts about business; the facts are usually not available and libel laws and mores greatly restrict their use. Even in Great Britain, hearings with a view to enacting antimonopoly legislation or revising the patent law, which necessarily expose sordid records, questionable practices, and caustic opinions, are not made available to the public. Newspapers and speakers are very hesitant and cautious in disseminating such information about business, however truthful it may be. "Would it hurt a company?" rather than, "Would it reveal the truth?" is the test. Consequently those who object to exploitation are greatly restricted in building up a forceful opinion against it. In all these respects the United States offers a complete contrast. This is the only major country of the world in which the soiled linen is hung upon the line for all to see. Facts about the wrong doing of business are available through printed hearings of Congress, government reports, and even court decisions. Any publisher, teacher, or politician may quote such material to the limit, interpret it, and demand remedial action. Note the exposé of the cartel arrangements concerning synthetic rubber. Where in Europe may one go to obtain information on cartels and, if it could be had, who would dare to launch a campaign against the participants? Though publicity itself is wholesome, remedial action may never come in this country because of our checks and balances—House, Senate, President, and the courts must approve a law before it becomes effective—and that is another story.

CHAPTER VI

Control of Supplementary Products

MANY BUSINESS ENTERPRISES have unduly extended their patent monopolies to products used with their inventions. These products have been either supplies and parts for a particular invention or a machine and other goods employed in a previous or subsequent operation. In other words, these products have been either auxiliary or complementary. Some supplementary products may be patented, others unpatented. The invention to which they are tied may be a process, machine, composition, or combination. Patent owners have derived profit from marketing the invention or the supplementary goods or both. In some instances the invention itself has been provided free of charge. The seller or lessor of an invention has employed different schemes to compel or encourage the purchaser or lessee to obtain supplementary goods from him—printed notice, written and implied contracts, threats, discriminations, and the like. As just suggested, the control over supplementary supplies has been exercised on the basis either of alleged privileges which patents confer or of economic pressure which patent ownership permits. How far patent owners may legally extend such control has depended upon the interpretation of patent rights by the courts and the explicit and implied restrictions upon them in the Sherman and Clayton acts. How far they may illegally exercise such control hinges upon their ingenuity in devising new and subtle schemes and the effectiveness of the government in law enforcement.

Apparently no monopoly or restraint is more appealing to the profit seeker than the control of supplementary products, especially of supplies for a particular invention. Such supplies are bought frequently, the outlay each time is usually small, any charge above the competitive level seems insignificant in each transaction, and the payment is in proportion to the amount of use. The tribute is like a sales tax—apparently so little to the one who pays it from time to time and yet so much to the one who collects it. Moreover, if the invention is a machine, it may be leased, so that residual control remains in the patent-licensor.

As one might expect, the judicial history of the control of supplementary supplies is long and varied. It may be divided roughly into four periods—uncertain status of the practice from the beginning of our patent system to 1896; qualified approval from 1896 to 1912; approval from 1912 to 1917; and disapproval to an increasing extent from 1917 to the present. The years 1896, 1912, and 1917—mark the beginnings of new trends of interpretation.

Uncertainty until 1896

The early decisions of the Supreme Court do not clearly and consistently reveal the extent to which the patentee may license another upon restrictive conditions. One court decision indicates that the patentee may market his invention upon the condition that the right to make, use, and sell shall not extend beyond a certain time and that any subsequent transfer is subject to this restriction. Another states that one cannot legally remake or reconstruct a patented invention from its used parts. Several of the court's early decisions indicate or imply that patent laws do not furnish a legal basis for controlling supplementary supplies. They clearly show that the purchaser of a patented invention has the right to repair it or replace unpatented parts.

Planing Machine and Knives. Wilson v. *Simpson* involved a combination planing machine with unpatented cutting knives or planes. The machine would last for several years but the unpatented knives would wear out within ninety days. The issue was whether an assignee had a right to replace new knives for those which wore out. The court in 1850 declared, "The right to replace them was a part of the invention transferred to the assignee for the time he bought it, without which his purchase would have been useless to him, except for sixty or ninety days after a machine had been put in use."[1] The replacement of parts, according to the court, is a repair, not a reconstruction of the invention. The court took pains to point out "The right of the assignee to replace the cutter-knives is not because they are

[1] *Wilson* v. *Simpson,* 9 Howard 109, 125 (1850).

of perishable materials, but because the inventor of the machine has so arranged them as a part of its combination, that the machine could not be continued in use without a succession of knives at short intervals."[2]

Rights to Patented Machine under Extension of Life of Patent. In 1852, the Supreme Court, in a leading case entitled *Bloomer* v. *McQuewan,* decided that a patentee, once having received his remuneration or royalty from his invention, cannot subsequently treat the buyer or user as an infringer. It held that a person who had purchased a patented machine, and was using it during the original term for which the patent was granted, might continue to use the machine during an extended term. "In using it, he exercises no rights created by the act of Congress, nor does he derive title to it by virtue of the franchise or exclusive privilege granted to the patentee. The inventor might lawfully sell it to him, whether he had a patent or not, if no other patentee stood in his way. And when the machine passes to the hands of the purchaser, it is no longer within the limit of the monopoly. It passes outside of it, and is no longer under the protection of the act of Congress."[3] Other decisions reflect this same view; for example, *Chaffee* v. *Boston Belting Co.*[4] and *Bloomer* v. *Millinger.*[5] This line of argument in these cases is the historical and judicial background for recent decisions of the Supreme Court which deny the right of the patent owner to dictate supplementary supplies or resale prices.

Reserved Right of Patentee in Sale of Invention. In *Mitchell* v. *Hawley,* a patentee licensed another to make, use, and sell a felting machine upon the condition that the right was not to extend beyond the original term of the patent and that any transfer of it was subject to this restriction. Upon the expiration of this term, there was a renewal of the patent but the purchaser of a machine from the licensee continued to use it. In response to a suit for infringement, the Supreme Court in 1872 held that the purchaser's title was no better than that of the seller—in fact, that this restriction by contract bound him as well as the original licensee.[6] Does this decision constitute a

[2] *Ibid.,* 125.
[3] *Bloomer* v. *McQuewan,* 14 How. 539, 549 (1852).
[4] 22 How. 217 (1859).
[5] I Wall. 340 (1863).
[6] *Mitchell* v. *Hawley,* 16 Wall. 544 (1872).

recognition of a limited license or is it explainable as being merely an overemphasis on contract? In 1912, in *Henry* v. *A. B. Dick,* this doctrine was construed as a legal opportunity for selling a patented invention upon the condition that supplementary supplies be bought from the patent owner.

Metallic Cotton-Bale Tie. The Cotton-Tie Company owned a patent on a metallic cotton-bale tie, consisting of a buckle and band, and sold the tie with these words stamped in metal: "Licensed to use only once." After the bands had been severed, the Simmons Company bought them as scrap iron, rolled and straightened the pieces, riveted together their ends, cut them into proper lengths, and sold them with the buckles, for use as ties. The Supreme Court, in 1882, decided that the single use of a tie, and its subsequent severance, were intended to operate as a destruction of it and that putting the same parts together for use a second time constituted an infringement. Therefore, the Simmons Company became an infringer not because of violating the "license to use only once" but because of reconstructing the bands after they had been cut—in other words, of remaking the invention after it had been destroyed.[7] A similar decision in 1950 held that putting new tips on patented oil-well drill bits constitutes infringement.[8]

Dispensing Machines and Paper. The Morgan Envelope Company manufactured and sold a fixture for dispensing paper. The fixture was patented while the paper was not. Morgan did not sell the fixture independently but only with the paper; however, there was no license restriction to this effect. The Albany Paper Company manufactured and sold paper to those who had previously bought fixtures and paper from Morgan; also, in some instances it bought these fixtures from original purchasers and sold them and paper for them, to others.

The Supreme Court, in 1894, said that the real question is, whether, conceding a valid combination of fixture and paper, "the sale of one element of such combination, with the intent that it shall be used with the other element, is an infringement. We are of opinion that it is not. There are doubtless many cases to the effect that the manufacture and sale of a single element of a combination, with intent that

[7] *Cotton-Tie Co.* v. *Simmons,* 106 U.S. 89, 91 (1882).

[8] *Williams* v. *Hughes Tool Co.,* 186 F. 2d 278 (10th Cir. 1950); affirming 82 F. Supp. 408 (D.C. Okla. 1949).

it shall be united to the other elements, and so complete the combination, is an infringement. But we think these cases have no application to one where the element made by the alleged infringer is an article of manufacture perishable in its nature, which it is the object of the mechanism to deliver, and which must be renewed periodically, whenever the device is put to use."[9] In short, the paper is transitory and passive, and its use does not reconstruct or recreate the invention. These physical characteristics of the product seemed important in delineating the rights of the patentee. In the decision there is no mention of the Sherman Act which had become a law only four years earlier.

Qualified Approval, 1896–1912

In 1896, a circuit court of appeals decided that a patentee could, by printed notice, control the supplies for his invention. This decision served as a precedent in other decisions of lower courts. However, the Supreme Court did not sanction this interpretation until 1912. This decision of 1896 was reached in the so-called button fastener case.

Fastening Machines and Fasteners. The Heaton-Peninsular Button Fastener Company owned several patents on a device for fastening buttons to shoes with metallic fasteners. A restriction inscribed upon a metal label was affixed to each machine:

> Condition of Sale. This machine is sold and purchased to use only with fasteners made by the Peninsular Novelty Company, to whom the title to said machine immediately reverts upon violation of this contract of sale.[10]

The fasteners or staples were not patented. The manufacturer of devices or machines sold them at cost to shoe dealers; it expected to receive its profits from the sale of the staples for them. The Eureka Specialty Company, knowing of the patents and notice of Peninsular, sold staples adapted for use with the latter's machine. Peninsular sued Eureka as a contributory infringer, as one who had co-operated in a use forbidden by the notice. In this case, in contrast with the previous one involving dispensing machines and paper, the unpatented article is used with the patented machine instead of being an element in a

[9] *Morgan Envelope Co.* v. *Albany Paper Co.,* 152 U.S. 425, 432–33 (1894).

[10] *Heaton-Peninsular Button-Fastener Co.* v. *Eureka Specialty Co.,* 77 Fed. 288, 290 (6th Cir. 1896).

combination, and the suit is against a contributory infringer—a fairly new type of offender at that time.

The court, in 1896, declared that the purchaser of the fastening machine acquired and accepted it "with a license to use the invention only with staples made by the patentee"[11] and that those who aided such infringement by intentionally and maliciously persuading and inducing the licensees to exceed their licenses by furnishing them with the means for such infringement were themselves infringers. Also it held that the sale of the machine through jobbers, and not directly to the users of it, was immaterial.

In its reasoning the court said that the buyer of the machine obtains the title to the materials embodying the invention, but only a license to use the invention. It may be properly asked how one may become the owner of the materials embodying an invention and not of the use of the invention. The materials are bought as a unified product for a particular use and not for the metal, wood, etc., which they represent. An economic good may be defined as the sum of its uses. Therefore, how can one be the owner of the matter which it contains and only a licensee to use it? The distinction is a nice bit of legal metaphysics.

In discussing the economic aspects of the case, the court said that the button fastening manufacturer chose "to fix the price for the right of use at the profit resulting from the sale of the staples," and that the extent of monopoly in unpatented staples will depend upon the merit of the patented fastener.

This decision was rendered by three judges two of whom—Horace H. Lurton and William H. Taft—later became members of the Supreme Court.

Talking Machines and Records. The Victor Talking Machine Company manufactured and sold a combination consisting of a patented talking machine and records. Leeds and Catlin Company made and sold records adapted for use on Victor talking machines. The Supreme Court, in 1909, held that these records, though unpatented, were an essential part of the patented combination of Victor and therefore that Leeds and Catlin was guilty of contributory infringement.[12] The court stated that an invention may be a combination of

[11] *Ibid.,* 291.

[12] *Leeds and Catlin* v. *Victor Talking Machine Co.,* 213 U.S. 325 (1909).

elements, some of which may be old and others new, or all old and all new. "It can make no difference as to the infringement or non-infringement of a combination that one of its elements or all of its elements are unpatented."[13]

The court contrasted this decision with the Albany Paper case, particularly the articles, records and paper respectively, used in combination with machines. The court described the records as active and durable, the paper as passive and perishable or transitory. It declared that the record or disc "co-acts"[14] with the machine to produce the result and that records are not "subject to decay by their inherent qualities, or consumed by few uses or a single use."[15] Leeds and Catlin, it held, knowingly sold to owners of Victor machines in order to increase the owners' stock of records rather than to repair or replace Victor records. As stated by the Supreme Court in a later decision, "The question was not of a right to substitute worn out parts but of a right to use new discs in a talking machine."[16] In other words the issue was not one of repair but of reconstruction of the invention.

Approval from 1912 to 1917

In 1912, for the first time—and also the last and only time—the Supreme Court went all the way in approving the right of a patentee to control, by printed notice, the sale of supplies and materials for his invention. This issue was raised in 1882 in the Cotton-Tie case and, in 1907, in *Cortelyou* v. *Johnson,*[17] and in both instances it was brushed aside by other considerations. As already noted, the Supreme Court settled the Cotton-Tie case on other grounds. In *Cortelyou* v. *Johnson,* as in the Dick case, described below, a duplicating machine was sold with the license restriction that it could be used only with stencil paper, ink, and other supplies made by the manufacturer. Johnson was charged with contributory infringement because it made unpatented ink and sold it to the purchasers of these machines. The Supreme Court held that Johnson did not have sufficient notice of the restriction; it did not pass upon the legality of the restriction itself.

[13] *Ibid.,* 333.
[14] *Ibid.,* 335.
[15] *Ibid.,* 335–36.
[16] *Heyer* v. *Duplicator Mfg. Co.,* 263 U.S. 100, 102 (1923).
[17] 207 U.S. 196 (1907).

Duplicating Machines, Stencils, Ink, and Other Supplies. The A. B. Dick Company, the manufacturer of a patented stencil duplicating machine, known as the "rotary mimeograph," sold the machine with a license restriction:

> This machine is sold by the A. B. Dick Company with the license restriction that it may be used only with the stencil paper, ink, and other supplies made by A. B. Dick Company, Chicago, U. S. A.[18]

The Henry Company, the defendant, had made a special ink designed for use on the mimeograph, and, with knowledge of the printed restriction, had sold this ink to an owner of one of Dick's machines. The Supreme Court declared him guilty of contributory infringement. Its decision in effect permitted the patentee to stipulate that he alone shall furnish the supplies used with his invention. As previously stated, lower federal courts had already upheld this interpretation in several instances.

After quoting *Mitchell* v. *Hawley* with approval, the Supreme Court pointed out that the greater right of exclusive use of the patentee embraces the lesser one of permitting the licensee to use an invention only upon prescribed conditions, and the court held that the patentee may sell his property right in the machine and at the same time may part with his property right in the monopoly covering the machine only upon certain conditions. In the court's own words, "This right to sever ownership and use is deducible from the nature of a patent monopoly."[19]

This reasoning is untenable from an economic standpoint, for, as a practical matter, how can the property right to a patented machine pass to a purchaser with no right to use? Would one buy the machine for any other purpose than to use it? The materials of which a mimeograph is built would cost much less than the mimeograph.

The court placed on the same plane the sundry restrictions upon the use of an invention. "There is no difference, in principle, between a sale subject to specific restrictions as to the time, place, or purpose of use, and restrictions requiring a use only with other things," and "If the violation of the one kind is an infringement, the other is also."[20] But is there not an important difference? In one instance the

[18] *Henry* v. *A. B. Dick Co.,* 224 U.S. 1, 11 (1912).

[19] *Ibid.,* 25.

[20] *Ibid.,* 35–36.

buyer can use the invention, though subject to a certain duration or area or use, independently of the seller; but in the other he cannot—he must seek out supplies from the seller only, regardless of inconvenience or price. Compare two purchases of stoves—one for use only in the home or for domestic purposes and one for use only with repairs or fuel obtained from the seller.

In this connection the court pointed out that the facts differed from those in the Albany Paper case. "First, because here the ink and other supplies used in the operation of the rotary mimeograph patent were not made elements of the patent, as in the Albany case; and secondly, the fixture in the Albany case was not sold with the license restriction that it was not to be used except in connection with paper supplied by the patentee."[21]

The court met the argument that the legality of such license restrictions might give the patentee the "practical monopoly of the market" for supplementary but unpatentable supplies, by saying that he alone created the market for such supplies by making and selling a new invention."[22] A fundamental weakness in this argument lies in the extent to which a patented machine or other article may be unpatented. All the patents covering it, except a few insignificant ones, may have expired. In this particular instance all the basic patents had terminated; but apparently this did not affect in the least the legality of the license restrictions as described. If improvement patents of slight importance permitted such restrictions, the use of tying clauses would become general because such patents are comparatively easy to obtain. In reply to the contention that such a restriction may be inconvenient and onerous, the court declared that "the public is always free to take or refuse the patented article on the terms imposed."[23] Does not the public have that much choice about any monopoly?

This decision of the Supreme Court, four to three, though agreeing with precedent as established by the lower courts, aroused considerable protest from the public. Unfortunately this majority opinion was handed down by only four of the nine judges, as follows: Lurton, who wrote the decision in the button fastener case; Van Devanter, in *National Phonograph Co.* v. *Schlegel;* McKenna, in the paper bag

[21] *Ibid.,* 48.
[22] *Ibid.,* 32.
[23] *Ibid.,* 34.

case; and Holmes, who dissented in the Dr. Miles case. These four judges dissented in the Sanatogen case. Chief Justice Edward D. White and Justices Charles Evans Hughes and Joseph R. Lamar joined in a very strong dissenting opinion. They maintained that the majority opinion contained the potentialities of dire consequence. Justice William Rufus Day, a new appointee, did not participate—there were only eight members at the time.

Partly in response to the Dick decision but largely because of the United Shoe Machinery monopoly, Congress passed the Clayton Act of 1914, section three of which provides that it is unlawful to lease or sell machinery, whether patented or unpatented, on the condition that the lessee or purchaser shall not use or deal in the machinery of a competitor if the effect "may be to substantially lessen competition or tend to create a monopoly in any line of commerce." In 1917, the Federal Trade Commission issued a complaint against the Dick Company to prevent its use of the tying clauses already described. In the same year, after the Supreme Court announced the projecting machine decision, Dick dropped its legal defense in the case, and the commission issued an order prohibiting Dick "from selling its machines and paper upon the condition, agreement, or understanding, whether embodied in a so-called license restriction or in a contract, or otherwise, that the purchaser should not use therewith the machines or supplies of a competitor."[24]

Disapproval to an Increasing Extent since 1917

The Supreme Court's four to three approval in 1912 of the right to dictate by printed notice the supplementary supplies for a patented invention was followed by its six to three disapproval and reversal in 1917. In 1913, the Supreme Court in *Bauer* v. *O'Donnell* rejected five to four, the basic propositions underlying *Henry* v. *Dick;* the four judges who made up the minority in the one case were the majority of four in the other. However, this tribunal did not definitely reverse *Henry* v. *Dick* until 1917, when the first test case provided the opportunity. Therefore according to the record, the patent owner, from 1912 to 1917, could and did legally control the supplementary products for his invention.

[24] Federal Trade Commission Annual *Report* (1917), 9.

The decision of 1917 brought the Supreme Court back again to its original line of reasoning in *Bloomer* v. *McQuewan* and other early cases. In 1944, the court declared that since 1917 it "has consistently held that the owner of a patent may not employ it to secure a limited monopoly of an unpatented material used in applying the invention."[25]

Projecting Machines and Films. The Motion Picture Patents Company owned a patent on a projecting machine and attempted, by printed notice, to define the films with which it must be used. The facts were substantially the same as those in the mimeograph case. The Universal Film Company made films and sold them in violation of this notice, and as a result Motion Picture Patents sued for infringement.

A circuit court of appeals pronounced this license notice in conflict with Section 3 of the Clayton Act.[26] However, the Supreme Court upon appeal, based its decision upon the patent law rather than the Clayton Act. Six of the judges held that the tying clause could not be justified by the patent law while three of them, in a minority decision, followed the line of argument employed in the Dick case. These three—Justices Holmes, McKenna, and Van Devanter—were also the three dissenters in the *Straus* v. *Victor* case; both cases were decided the same day. The court held that in interpreting the patent law, it would be of service to keep in mind these three rules: the "scope of every patent is limited to the invention described in the claims contained in it, read in the light of the specification"; the "patentee receives nothing from the law which he did not have before," and the "only effect of his patent is to restrain others from manufacturing, using, or selling that which he had invented"; and the "primary purpose of our patent laws is not the creation of private fortune for the owners of patents, but is 'to promote the progress of science and useful arts.' "[27]

In view of these rules the court decided that the patent granted on a machine is restricted "to the mechanism described in the patent as necessary to produce the described results. It is not concerned with and has nothing to do with the materials with which or on which the

[25] *Mercoid Corp.* v. *Mid-Continent Investment Co.,* 320 U.S. 661, 664 (1944).

[26] 235 Fed. 398 (2d Cir. 1916).

[27] *Motion Picture Patents Co.* v. *Universal Film Mfg. Co.,* 243 U.S. 502, 510–11 (1917).

machine operates."[28] For, "whatever right the owner may have to control by restriction the materials to be used in operating the machine must be derived through the general law from the ownership of the property in the machine, and it cannot be derived from or protected by the patent law, which allows a grant only of the right to an exclusive use of the new and useful discovery which has been made —this and nothing more."[29] Note how this statement leaves open the question whether such a restriction may be imposed by contract.

The economic evils which could arise from license restrictions like the one in question received consideration in the decision: "The perfect instrument of favoritism and oppression which such a system of doing business, if valid, would put into the control of the owner of such a patent, should make courts astute, if need be, to defeat its operation."[30] The court gave an effective answer to the argument that the license restriction redounded to the economic welfare of the public. "It is argued as a merit of this system of sale under a license notice that the public is benefited by the sale of the machine at what is practically its cost, and by the fact that the owner of the patent makes its entire profit from the sale of the supplies with which it is operated. This fact, if it be a fact, instead of commending, is the clearest possible condemnation of, the practice adopted, for it proves that under color of its patent the owner intends to and does derive its profit, not from the invention on which the law gives it a monopoly, but from the unpatented supplies with which it is used, and which are wholly without the scope of the patent monopoly, thus in effect extending the power to the owner of the patent to fix the price to the public of the unpatented supplies as effectively as he may fix the price on the patented machine."[31] That this company had forty thousand projecting machines in use in the United States and that the mechanism covered by the patent in suit was the only one with which the motion-picture films could be used successfully were admitted facts which supported such condemnation by the court.

"It is obvious," the court said, "that the conclusions arrived at in this opinion are such that the decision in *Henry* v. *Dick* must be

[28] *Ibid.*, 512.
[29] *Ibid.*, 513.
[30] *Ibid.*, 515.
[31] *Ibid.*, 516–17

regarded as overruled."[32] The court thus pronounced the illegality of dictating supplementary supplies for patented articles. Unfortunately, however, as the remainder of this chapter indicates, the decision did not bring the trade practice to a close.

Containers and Dry Ice. The American Patents Development Corporation had a patent on a refrigerating transportation package. The Dry Ice Corporation was the exclusive licensee; its sole business was the manufacture of solid carbon dioxide which it sold under the name of Dry Ice. The Dry Ice Corporation did not make or sell transportation packages in which solid carbon dioxide was used as a refrigerant. It did not license others to make such packages on a royalty basis. It did not formally license buyers of its dry ice to use the invention in suit. But each of its invoices for dry ice bore a notice, in effect, that the patented container could be used only with dry ice purchased from it. The Carbice Corporation, another manufacturer of solid carbon dioxide, sold its product to customers of Dry Ice Corporation with knowledge that it would be used by the purchaser in packages like those described in the patent. It challenged the validity of the patent and denied the alleged infringement.

The Supreme Court held unanimously that the Carbice Corporation had not invaded any right of the patentee.[33] It stated that the Dry Ice Corporation "has no right to be free from competition in the sale of solid carbon dioxide. Control over the supply of such unpatented material is beyond the scope of the patentee's monopoly; and this limitation, inherent in the patent grant, is not dependent upon the peculiar function or character of the unpatented material or on the way in which it is used. Relief is denied because the Dry Ice Corporation is attempting, without sanction of law, to employ the patent to secure a limited monopoly of unpatented material used in applying the invention."[34]

The court differentiated this case from *Leeds and Catlin* v. *Victor Talking Machine Co.* "That was an ordinary case of contributory infringement. The Victor company sold machines embodying a patent for a combination. Leeds and Catlin were held to be infringers because the intended incorporation in the Victor machines of the article which

[32] *Ibid.*, 518.
[33] *Carbice Corp.* v. *American Patents Development Corp.*, 283 U.S. 27 (1931).
[34] *Ibid.*, 33–34.

they sold, did not constitute a repair of the machine and hence was not within the license implied on sale. There was no suggestion that the Victor company, which itself manufactured and sold the patented product, sought 'to derive its profits, not from the invention on which the law gives it a monopoly, but from the unpatented supplies with which it is used.' In the case at bar the plaintiffs neither sell nor license others to sell complete transportation packages. They supply merely one of the several materials entering into the combination; and on that commodity, they have not been granted a monopoly. Their attempt to secure one cannot be sanctioned."[35] A few weeks after this decision the Supreme Court appraised the patent on the refrigerating transportation package and held it invalid for lack of novelty and invention.[36]

In thus attempting to reconcile these two decisions, twenty-two years apart, the Supreme Court accepted both of them—a fact which illustrates the unwillingness of courts to overrule prior cases unless absolutely necessary. These and kindred decisions reveal fine distinctions that have been used by this tribunal in interpreting the right of a patentee to dictate supplementary supplies: Does the patentee sell only the materials in his product or his entire patented invention? In other words, is the sale conditional or unconditional? Is the supply in question a part of a patented combination or only a material for the operation of the invention? Does the patentee derive his profit from his invention or from his supplies? Is the supply durable or transitory and perishable? Is it active or passive? Is it, in effect, for reconstruction or mere repair of the invention?

The judicial decisions considered thus far indicate that if the patent owner sells conditionally, if the supply is a part of a patented combination, if he secures his profit from the supplies, and if the supply is durable and active and for reconstruction of the invention, he may require the users of his invention to obtain supplementary goods only from him. On the other hand, if he parts with his invention unconditionally; sells one thing rather than an entire combination; if he obtains his profit from his invention rather than the supplies for it; if the supply is transitory rather than durable, passive rather than active, for repair rather than reconstruction of the invention,

35 *Ibid.*, 34–35.
36 *Ibid.*, 283 U.S. 420 (1931).

he cannot compel users to purchase necessary supplies only from him.

As will be noted further on, later decisions of the Supreme Court apparently discarded these distinctions and unreservedly denied the right of a patentee to control unpatented supplies of any sort.

Tabulating Machines and Cards. In 1936, International Business Machines Corporation and Remington Rand were the only manufacturers of machines "which perform certain mechanical tabulations and computations" by the use of cards upon which data are recorded.[37] Each of the two companies leased its machines upon the condition that the lessees use with the machines only cards which the manufacturer of the machine provided. IBM contended that this condition did not extend beyond the monopoly which it had acquired by separate patents on its cards and on its machines. After stating that IBM's patents on cards appear to extend only to the cards when perforated, the Supreme Court declared, "But we do not place our decision on this narrow ground. We rest it rather in the language of Section 3 of the Clayton Act which expressly makes tying clauses unlawful, whether the machine leased is 'patented or unpatented.' "[38] The court held that this condition or tying clause operated to prevent competition and to create a monopoly in the production and sale of tabulating cards suitable for the machines. In this connection it pointed out that IBM made 81 per cent of these cards—Remington the remainder—from which IBM's gross receipts during the previous ten years had averaged $3,192,700 annually.[39]

In reply to the contention that the manufacturer of the machines should furnish the cards so that they will operate successfully, the court said that evidence indicates that others can manufacture cards suitable for use in these machines; for example, the government, under special provisions of its lease, had made large quantities of cards by its own methods and used them successfully with these machines. In conclusion, however, the court agreed that IBM "is not prevented from proclaiming the virtues of its own cards or warning against the danger of using, in its machines, cards which do not conform to the necessary specifications, or even from making its leases conditional upon the use of cards which conform to them. For aught that appears

[37] *International Business Machines Corp.* v. *United States,* 298 U.S. 131 (1936).
[38] *Ibid.,* 137.
[39] *Ibid.,* 136.

such measures would protect its good will, without the creation of monopoly or resort to the suppression of competition."[40]

Bituminous Emulsion. Various companies manufactured a bituminous emulsion, an unpatented staple article of commerce used by their customers for many purposes. "With builders of cement concrete roads it has recently come into use for a film on the surface of the roadway to retard evaporation during curing. For the method of so retarding evaporation the Barber Company acquired the process patent sued on, and seeks to use it to secure a limited monopoly in the business of producing and selling the bituminous material for practicing and carrying out the patented method. The company does not itself engage in road building or compete with road contractors. It does not seek to make road builders pay a royalty for employing the patented method. It does not grant to road builders a written license to use this process. But it adopts a method of doing the business which is the practical equivalent of granting a written license with a condition that the patented method may be practiced only with the emulsion purchased from it. For any road builder can buy emulsion from it for that purpose, and whenever such a sale is made, the law implies authority to practice the invention. On the other hand the Barber Company sues as contributory infringer a competing manufacturer of this unpatented material who sells it to a road builder for such use. Thus, the sole purpose to which the patent is put is thereby to suppress competition in the production and sale of staple unpatented material for this use in road building."[41]

Barber maintained that the Carbice decision was not applicable because Barber did not employ any contract or notice to expand its patent monopoly. The court held that such a distinction has no legal significance. Referring to the Carbice case it said, "By the rule there declared, every use of a patent as a means of obtaining a limited monopoly of unpatented material is prohibited. It applies whether the patent be for a machine, a product, or a process. It applies whatever the nature of the device by which the owner of the patent seeks to effect such unauthorized extension of the monopoly. Nothing in *Leeds and Catlin Co.* v. *Victor Talking Machine Co.* limits it."[42] In this

40 *Ibid.*, 139–40.
41 *Leitch Mfg. Co.* v. *Barber Co.*, 302 U.S. 458, 460 (1938).
42 *Ibid.*, 463.

case the court applies the misuse doctrine in a suit against an infringer, rather than in one against a licensee, and apparently condemns all variations of restrictive controls upon unpatented supplies by the patent owner.

Salt Tablets. The Morton Salt Company had a patent on a device used by canners for adding predetermined amounts of salt to the contents of cans. The principal business of Morton's wholly-owned subsidiary, from which Morton's profits were derived, was the manufacture and sale of salt tablets. Morton leased its patented machines to canners, about two hundred in number, upon the condition that only its salt tablets be used in the machines' operation. The Suppiger Company made and leased to canners unpatented salt-depositing machines and was charged with infringement. Nothing in the resulting decision turned on the fact that Suppiger also competed with Morton in the sale of salt tablets. The Supreme Court stated that the use of a patent "to suppress competition in the sale of an unpatented article may deprive the patentee of the aid of a court of equity to restrain an alleged infringement by one who is a competitor."[43] In effect, it refused to consider Morton's charge of alleged infringement of its machine because of Morton's restraint of competition in the sale of unpatented salt, and thus the court added another hazard to the patent owner who seeks to control the supplementary supplies for his invention.

Adhesives and Applying Machines. The same day the Supreme Court handed down the previous decision, it rendered a similar one which involved the owner of a patent on a method of using fabric for reinforcing insoles in shoe manufacture.[44] The patent owner supplied adhesive and a patented machine for applying it to the fabric, and charged each manufacturer according to the amount of fabric used. B. B. Chemical Company sold material to the shoe manufacturers for use by this patented method and instructed them as to its use, whereupon the patent owner sued for infringement.

The Supreme Court said that this case, analogous to the last one, "involves the question whether the owner of a method patent who authorizes manufacturers to use it only with materials furnished by him may enjoin infringement by one who supplies the manufacturer

[43] *Morton Salt Co.* v. *Suppiger Co.,* 314 U.S. 488, 491 (1942).
[44] *B. B. Chemical Co.* v. *Ellis,* 314 U.S. 495 (1942).

with materials for use by the patented method and aids in such use."[45] It held that even though the infringement by the chemical company did extend beyond the mere sale of the materials to the manufacturers, "the maintenance of this suit to restrain any form of infringement is contrary to public policy."[46]

In discussing the last two cases, the Supreme Court in the Hartford-Empire case said that they "merely apply the doctrine that, so long as the patent owner is using his patent in violation of the antitrust laws, he cannot restrain infringement of it by others. We were not there concerned with the problem whether, when a violation of the antitrust laws was to be restrained and discontinued, the court could, as part of the relief, forfeit the patents of those who had been guilty of the violation."[47] But if a patent is unenforceable in one instance because of misuse, why should it not be unenforceable in another?

Domestic Heating Systems and Combination Stoker Switches. The Mid-Continent Investment Company owned the Cross combination patent for a domestic heating system which comprised three main elements—a motor-driven stoker for feeding fuel to the combustion chamber of a furnace, a room thermostat for controlling the feeding of fuel, and a combustion stoker switch to prevent extinguishment of the fire.[48] Mid-Continent Investment granted the Minneapolis-Honeywell Regulator Company an exclusive license to make, use, and sell and to sublicense others. The royalty payments under the license were based only upon the sales of the combustion stoker switch which was an element of the combination patent embodied in the patented article but which was itself unpatented. Minneapolis-Honeywell in advertising its stoker switches stated that the "right to use" the Cross system patent was "only granted to the user" when Minneapolis-Honeywell stoker switches were purchased and used in the system. These two companies—Mid-Continent Investment and Minneapolis-Honeywell—neither manufactured nor installed heating systems under the patent; nevertheless they endeavored to use the license agreement so as to prevent the sale or use of the necessary switches unless made by Minneapolis-Honeywell and purchased from it or its sub-

[45] *Ibid.*, 495–96.

[46] *Ibid.*, 498.

[47] *Hartford-Empire Co.* v. *United States,* 323 U.S. 386, 415 (1945).

[48] *Mercoid Corp.* v. *Mid-Continent Investment Co.,* 320 U.S. 661, 664 (1944).

licensees.[49] The Mercoid Corporation likewise did not sell or install the Cross heating system, but it knowingly manufactured and sold combustion stoker switches for use in this system; as a result it was charged with contributory infringement.[50]

The Supreme Court stated that ever since the mimeograph decision was overruled in the projecting machine case the court "has consistently held that the owner of a patent may not employ it to secure a limited monopoly of an unpatented material used in applying the invention."[51] Later and kindred cases, already noted, "involved the use of the patent for a machine or process to secure a partial monopoly in supplies consumed in its operation or unpatented materials employed in it. But we can see no difference in principle where the unpatented material or device is itself an integral part of the structure embodying the patent."[52]

With these words the Supreme Court in effect overruled its decision in *Leeds and Catlin* v. *Victor*—a decision which "is authority for the conclusion that he who sells an unpatented part of a combination patent for use in the assembled machine may be guilty of contributory infringement. The protection which the court in that case extended to the phonograph record, which was an unpatented part of the patented phonograph, is in substance inconsistent with the view which we have expressed in this case."[53] It should be added that the court also rejected the language in the Carbice case in which it attempts to reconcile the Carbice and the Leeds and Catlin opinions.

In a similar case decided the same day as *Mercoid Corp.* v. *Mid-Continent Investment Co.*, the Supreme Court said, "However worthy it may be, however essential to the patent, an unpatented part of a combination patent is no more entitled to monopolistic protection than any other unpatented device."[54] Apparently the court thus refused to sanction the extension of the patent monopoly to any unpatented article under any circumstance. At the same time, therefore, it pushed aside the fine distinctions of earlier cases—profit from invention *v.* profit from supplies, durability *v.* perishability, active *v.* passive, etc.,

49 *Ibid.*, 663.
50 *Ibid.*, 664.
51 *Ibid.*, 664.
52 *Ibid.*, 665.
53 *Ibid.*, 668.
54 *Mercoid Corp.* v. *Minneapolis-Honeywell Regulator Co.*, 320 U.S. 680, 684 (1944).

as well as part of a combination *v.* only a material—all of which are enumerated after the discussion of the Carbice case.

The court declared, "The legality of any attempt to bring unpatented goods within the protection of the patent is measured by the antitrust laws—not by the patent law."[55] Compare this with its point of view forty-two years earlier: "The general rule is absolute freedom in the use or sale of patent rights under the patent laws of the United States."[56]

Dissolving and Injecting Machines and Salt. The International Salt Company, the country's largest producer of salt for industrial uses, manufactured and distributed two patented machines along with unpatented salt. One of the machines dissolved rock salt into a brine, and the other injected a salt tablet into a container. International generally leased these machines for nominal amounts upon condition that the licensees purchase the necessary salt from it.[57]

These facts are similar to those in the Morton Salt case. However, in this instance the issue was the violation of the Sherman and Clayton laws rather than infringement. In 1947, the Supreme Court declared, "By contracting to close this market for salt against competition, International has engaged in a restraint of trade for which its patents afford no immunity from the antitrust laws."[58] The court went on to say that "It is unreasonable, per se, to foreclose competitors from any substantial market."[59] Also, under the Clayton Act, "agreements are forbidden which 'tend to create a monopoly,' and it is immaterial that the tendency is a creeping one rather than one that proceeds at full gallop; nor does the law await arrival at the goal before condemning the direction of the movement."[60]

The contracts were not saved from illegality because they permitted lessees to buy salt on the open market if any competitor offered it of equal grade at a lower price and if International did not meet the price. This "provision does, of course, afford a measure of protection to the lessee, but it does not avoid the stifling effect" of a "priority on the business at equal prices."[61]

[55] *Ibid.*, 684.
[56] *Bement* v. *National Harrow Co.*, 186 U.S. 70, 91 (1902).
[57] *International Salt Co.* v. *United States*, 332 U.S. 392, 394 (1947).
[58] *Ibid.*, 396.
[59] *Ibid.*, 396.
[60] *Ibid.*, 396.
[61] *Ibid.*, 397.

The court conceded that a lessor may impose restrictions to assure satisfactory operation; for example, if the machine functioned best on rock salt 98.2 per cent pure—a percentage claimed by International for its salt—the lessee might be required to use only salt "meeting such a specification of quality. But it is not pleaded, nor is it argued, that the machine is allergic to salt of equal quality produced by anyone except International."[62] The court concluded, "Rules for use of leased machinery must not be disguised restraints of free competition, though they may set reasonable standards which all suppliers must meet."[63]

Another significant aspect of this case is the remedy as prescribed by the lower court and approved with modification by the Supreme Court six to three—Justices Vinson, Black, Douglas, Murphy, Jackson, and Rutledge against Justices Reed, Frankfurter, and Burton. This remedy required International to sell or lease its machines to all applicants on nondiscriminatory terms and permitted the lower court to retain jurisdiction to consider applications for changes in its decree. International objected to this remedy. It contended that the mere decision which invalidated "the objectionable leases should end the matter" and therefore that otherwise it should be "entitled to stand before the court in the same position as one who had never violated the law at all."[64] However, the Supreme Court pointed out that International already had wedged itself into the salt market by methods forbidden by law. "And advantages already in hand may be held by methods more subtle and informed, and more difficult to prove, than those which, in the first place, win a market. When the purpose to restrain trade appears from a clear violation of law, it is not necessary that all of the untraveled roads to that end be left open and that only the worn one be closed. The usual ways to the prohibited goal may be blocked against the proven transgressor and the burden put upon him to bring any proper claims for relief to the court's attention."[65]

After pointing out that antitrust suits should effectively pry open to competition a market that has been closed by illegal restraints, the Supreme Court stated, "If this decree accomplished less than

[62] *Ibid.*, 398.
[63] *Ibid.*, 398.
[64] *Ibid.*, 400.
[65] *Ibid.*, 400.

that, the Government has won a lawsuit and lost a cause."[66] In a footnote it added that a district court is authorized, but not required, to call upon the Federal Trade Commission to assist in framing decrees in antitrust cases. This would "seem unnecessary if Congress intended a simple prohibition of the particular practice proved before the court. It indicates the Congress has intended decrees to deal with the future economic condition of the enterprise as well as past violations."[67]

The minority opinion stated, "Inasmuch as the holder of patents on machines is not obliged to dispose of them to all comers or to do so at a uniform price," the decree deprived International "of a legal right."[68]

This decree, along with similar ones beginning in 1942, indicates more strongly an intention to make the antitrust laws effective—to restore competition as well as to win a lawsuit. Too many times those found guilty of monopoly and restraint of trade still obtain their objectives by other methods.

Cans. The American Can Company sought to monopolize the tin can industry by acquiring control of the manufacture of cans as well as can-making machinery. In 1916, a district court held that it was a monopoly but that dissolution would be too drastic.[69] In 1924, as a result of an extensive hearing before the Federal Trade Commission, this company agreed to remove the tying clause from its leases which required that only its cans be used with its machines. Subsequently, it employed long-term requirement contracts with quantity discounts which served "as an inducement for canners to purchase all their needs from a single manufacturer."[70] In November, 1949, the court held that American Can had violated both the Sherman and Clayton acts. A litigated decree of June, 1950, forbade long-term requirement contracts and quantity discounts and ordered the leasing of closing machines to canners and the servicing of them regardless of the source of cans, and the sale of machines to any canner wishing to purchase them. It ordered royalty-free licensing of all present patents on closing machines; unrestricted licensing of future patents for five years;

66 *Ibid.*, 401.
67 *Ibid.*, 400–401.
68 *Ibid.*, 403.
69 *United States* v. *American Can Co.*, 230 Fed. 859, 901 (D. Md. 1916).
70 *Ibid.*, 87 F. Supp. 18, 24 (N.D. Calif. 1949).

and availability of know-how through methods, written information, consultation, observation in shops, and attendance in training schools —all either free or at actual cost. However, in 1954, the government charged American Can with violation of this decree by attempting tie-in sales of cans with leasing of closing machinery and by attempting discriminatory treatment.[71]

Shoe Machinery and Complementary Machines and Supplies. As indicated in Chapter Three, United Shoe Machinery manufactured and marketed a full line of machines and supplies with the exception of upper-fitting machinery after 1920. Shoe manufacturers as a class objected to United's full line forcing, but not to its leasing system. The royalties were the same for all manufacturers, whether they used a thousand machines or one; they were charged by the pair; and the service was uniform. The leasing system tended to make the expense of shoe manufacturing independent of volume and reduced the amount of capital necessary to engage in the business. Undoubtedly an underlying purpose of such a policy was to discourage the dominance of any shoe company and to discourage anybody's fabrication of shoe machinery.

The government, in 1911, specifically attacked these tying clauses as violative of the Sherman Act, but the Supreme Court in 1918, by a vote of four to three, upheld their legality.[72] After the passage of the Clayton Act in 1914, the government tested again the legality of these clauses, and the Supreme Court in 1922 declared them unlawful.[73] However, it made clear that these clauses were illegal because they violated the Clayton Act rather than the Sherman Act.

In 1953, Judge D. J. Wyzanski held that United's leases "have created barriers to the entry by competitors into the shoe machinery field"[74] for several reasons. "The lessee is now held closely to United by the combined effect of the ten-year term, the requirement that if he has work available he must use the machine to full capacity, and by the return charge which can in practice, through the right of deduction fund, be reduced to insignificance if he keeps this and

[71] *Ibid.*, CCH 1950–51 Trade Cases ¶ 62,679 (N.D. Calif. Civil No. 26345–H; Complaint, 1946; Final Decree, 1950); *ibid.* (N.D. Calif. Cr. No. 34089; Complaint, 1954).

[72] *United States* v. *United Shoe Machinery Corp.*, 247 U.S. 32 (1918).

[73] *United Shoe Machinery Corp.* v. *United States*, 258 U.S. 451 (1922).

[74] *United States* v. *United Shoe Machinery Corp.*, 110 F. Supp. 295, 340 (D. Mass. 1953).

other United machines to the end of the periods for which he leased them. Second, when a lessee desires to replace a United machine, United gives him more favorable terms if the replacement is by another United machine than if it is by a competitive machine. Third, United's practice of offering to repair, without separate charges, its leased machines, has had the effect that there are no independent service organizations to repair complicated machines."[75] It was the judge's opinion that "through its success with its principal and more complicated machines, United has been able to market more successfully its other machines."[76] In addition, "United has followed, as between machine types, a discriminatory pricing policy."[77] United receives a "higher rate of return where competition is of minor significance, and a lower rate of return where competition is of major significance."[78] The practice of leasing means "there is no substantial substitute competition from a vigorous second-hand market in shoe machinery."[79]

Other Products. a. *Hydraulic Storage System.* Complaints and consent decrees, in addition to the decisions already noted, reveal the control some patent owners maintain over supplementary products. One instance involved hydraulic gasoline-storage systems for aircraft, the building of which requires tanks, hose, cement, valves, meters, and the like. Aqua Systems and Flotation Systems, both incorporated, had patents covering some but not all of these parts and materials. The two systems cross-licensed each other and on numerous occasions refused to license other contractors under their combination patents required by specifications unless the latter purchased from them unpatented supplies at exorbitant prices. A consent decree in 1942 required the dedication to the public of eighteen patents and the grant of unrestricted royalty-free licenses under any subsequent patent issued on a pending application relating to hydraulic gasoline storage.[80]

b. *Vacuum-closing Machinery and Glass Containers.* Owens-Illi-

75 *Ibid.*, 340.

76 *Ibid.*, 340.

77 *Ibid.*, 340.

78 *Ibid.*, 341.

79 *Ibid.*, 344.

80 *United States* v. *Aqua Systems,* CCH Trade Regulation Reports, Supp. 1941–43, ¶ 52,880 (S.D.N.Y. Civil No. 19-516; Complaint, 1942; Consent Decrees, 1942).

nois Glass manufactured patented machines for creating a vacuum in glass containers and sealing them airtight, and beginning about 1935 leased these machines at a nominal rental for use only with Owens-Illinois containers. Owens-Illinois made and leased over 75 per cent of such machines and sold over 60 per cent of such containers. In September, 1946, a consent judgment, in response to government action the previous April, required Owens-Illinois to discontinue this restrictive practice and to license its patents on the machines at reasonable royalties—and in case of violation of this judgment, royalty-free.[81]

c. *Railway Springs and Spring Plates.* Symington-Gould Corporation and Universal Railway Devices Company, owners of patents on railway springs and plates, granted exclusive licenses to American Locomotive. This licensee in turn granted sublicenses to Baldwin Locomotive, Crucible Steel, and five other manufacturers of such products. Symington-Gould acted as the agent for getting orders for these springs and plates from railways and others. Upon receipt of orders, American Locomotive would distribute them to itself and its sublicensees, maintain uniform sales prices, compel railways and other customers to purchase unpatented castings and other supplementary parts from these manufacturers—all in keeping with agreements in 1932 and 1934. By 1942, American Locomotive and the sublicensees produced and sold over 90 per cent of the railway elliptical and coil springs and a substantial portion of the spring plates. Consent judgments in 1947 enjoined these restrictive practices, ordered Symington-Gould to assign claims under eleven patents to the public, and ordered Universal Railway Devices to license its patents and spring plates, until 1947, at reasonable royalties.[82]

d. *Axles and Axle Forgings.* The Timken-Detroit Axle Company owned or controlled patents relating to multi-wheel units for use on trucks and busses. These units consisted of sets of transversely-aligned wheels, so designed that each set supported a portion of the load at all times. The multi-wheel units were essential in the manufacture of large trucks and busses designed to carry heavy loads. At first, beginning in 1933, Timken-Detroit's license agreements with various manufacturers of trucks and busses required the manufac-

[81] *United States* v. *Owens-Illinois Glass Co.*, CCH 1946–47 Trade Cases ¶ 57,498 (N.D. Calif. Civil No. 25861–G; Complaint, 1946; Consent Judgment, 1946).

[82] *United States* v. *American Locomotive Co.*, CCH 1945–47 Trade Cases ¶ 57,621 (N.D. Ind. Civil No. 545; Complaint, 1945; Consent Judgment, 1947).

turers to purchase from Timken-Detroit the axles and axle forgings for use in all multi-wheel units. Timken-Detroit terminated these agreements in 1945—perhaps because of court decisions and decrees against such restrictions—and offered a new license without any tying clause but at a substantially higher rate of royalty. However, at practically the same time, Timken-Detroit informed manufacturers that they would be considered licensed under the patents without royalty if they would purchase exclusively from Timken-Detroit the axles and axle forgings to be used with multi-wheel units—an arrangement which would avoid the appearance but accomplish the effect of the former restrictions. The purchasers chose this arrangement. A consent decree in 1947 provided that any applicant might obtain a license under certain relevant patents upon payment of reasonable royalties.[83]

e. *Automatic Fire-extinguishing System.* Several automatic sprinkler companies, together with four subsidiaries, manufactured and distributed automatic fire-extinguishing systems. A part of the system was a patented "rate-of-rise" device which, when subjected to an abnormally rapid rise of temperature, caused the system to operate. Unpatented parts included pipes and sprinklers. Automatic Sprinkler, the only manufacturer of this patented device, sold the device only to certain companies and they in turn agreed to purchase supplementary equipment from Automatic. The defendants and their exclusive distributors installed more than 75 per cent of all rate-of-rise systems and followed uniform price lists. A consent judgment in 1948 required Automatic to sell its rate-of-rise devices to any prospective purchaser without discrimination and to license all applicants on an unconditional reasonable-royalty basis under all patents which it had or might acquire up to 1953.[84]

f. *Fastening Machines and Fasteners—Again.* The Universal Button Fastening and Button Company leased its patented machines for attaching fasteners to clothing, on condition that the machines be used only with Universal fasteners.[85] Almost identical cases involved

[83] *United States* v. *Timken-Detroit Axle Co.,* CCH 1946–47 Trade Cases ¶ 57,603 (E.D. Mich. Civil No. 5642; Complaint, 1946; Consent Judgment, 1947).

[84] *United States* v. *Automatic Sprinkler Co.,* CCH 1948–49 Trade Cases ¶ 62,230 (N.D. Ill. Civil No. 46-C-1289; Complaint, 1946; Consent Judgment, 1948).

[85] *United States* v. *Universal Button Fastening and Button Co.,* CCH 1948–49 Trade Cases ¶ 62,255 (E.D. Mich. Civil No. 5860; Complaint, 1946; Consent Judgment, 1948).

the Patent Button Company[86] and Scovill Manufacturing Company. The Scovill company attempted to justify its restriction on the ground that "said machine was designed specially for use in connection with the products manufactured by the lessor, and it is not intended to operate successfully with any other product," and then cautiously declared that "nothing herein shall be construed to limit in any way the lessee's right to acquire other machines or the product of any other manufacturer in connection with its business."[87] Consent decrees against these three companies, which enjoined the tying clauses and required patent licenses at reasonable royalties, came after twenty years of alleged violation of the Sherman and Clayton acts.

g. *Sealing Machinery and Caps.* The White Cap Company leased its sealing machines for food products on condition that the machines be used only with its caps. This company supplied 85 per cent of the vacuum caps used in packaging vegetables, fruits, coffee, and other food products. From 1926 to 1940, the lease agreements simply provided that "the lessee agrees to use said machine for the capping of containers with closure caps supplied to the lessee by the owner only." Apparently alarmed about government suits and consent decrees against such a restriction, White Cap, from 1940 to 1942, injected the phrase "as a safeguard against injury to said machine." A more ingenious disguise, beginning in 1942, stipulated that the owner of the machine would protect the lessee from all claims of infringement if the owner furnished the caps. A consent judgment in 1948 cancelled these agreements and required White Cap to license its machine patents to all applicants at reasonable rates.[88]

h. *Miscellaneous Group.* Standard Register Company leased and sold form-handling devices for use with business machines on condition that the devices be used only with the forms Standard manufactured. A consent judgment in 1949 enjoined this practice and required the licensing of Standard's patents at reasonable royalties.[89] Consolidated Car-Heating sold its dental alloy and leased its electric-

[86] *United States* v. *Patent Button Co.,* CCH 1946–47 Trade Cases ¶ 57,579 (D.C. Conn. Civil No. 1854; Complaint, 1946; Consent Judgment, 1947).

[87] *United States* v. *Scovill Mfg. Co.,* CCH 1948–49 Trade Cases ¶ 62,223 (D.C. Conn. Civil No. 1853; Complaint, 1946; Consent Judgment, 1948).

[88] *United States* v. *White Cap Co.,* CCH 1948–49 Trade Cases ¶ 62,268 (N.D. Ill. Civil No. 46–C–861; Complaint, 1946; Consent Judgment, 1948).

[89] *United States* v. *Standard Register Co.,* CCH 1950–51 Trade Cases ¶ 62,533 (D.C. of D.C. Civil No. 36040; Complaint, 1946; Consent Judgment, 1949).

casting machine for making dentures from metal alloys, on condition that the alloy and the machine be used only with Consolidated supplies. Judgments in 1950 and 1951 enjoined this practice.[90] Two companies—Parke Davis and Eli Lilly—leased capsule-filling machinery on condition that capsule requirements be purchased from them. A consent judgment in 1951 required the two companies to sell their machines, to license present patents royalty-free and future patents at reasonable royalties, and to furnish technical information.[91]

Companies described in previous chapters engaged in this practice until they were stopped by decisions or consent decrees. They include American Lecithin, Vehicular Parking, Parker Rust Proof, American Air Filter, and Bausch and Lomb and four other manufacturers of optical goods. Also, as previously indicated, Hartford-Empire, like United Shoe Machinery, used full line forcing in leasing its machines—an effective strategy because of its complete control of machines for some stages of manufacture.

Recent Legislation

In the Patent Act of 1952, Section 271, there appears a new definition of contributory infringement: "Whoever sells a component of a patented machine, manufacture, combination or composition, or a material or apparatus for use in practicing a patented process, constituting a material part of the invention, knowing the same to be especially made or especially adapted for use in an infringement of such patent, and not a staple article or commodity of commerce suitable for substantial manufacturing use, shall be liable as a contributory infringer." Would tabulating cards or salt tablets of certain dimensions and designs and therefore especially made or adapted for use on patented machines constitute contributory infringement? Would a rod sold for use only in the repair of a patented machine manufactured mostly or entirely by one company and therefore not suitable for substantial noninfringing use be contrary to this law? If the answers to these questions are in the affirmative, then important decisions of the Supreme Court which condemn the control of supplementary products

[90] *United States* v. *Consolidated Car-Heating Co.*, CCH 1950–51 Trade Cases ¶ 62,656 (S.D.N.Y. Civil No. 34–312; Complaint, 1945; Summary Judgment, 1950).

[91] *United States* v. *Parke, Davis and Co. and Eli Lilly and Co.*, CCH 1950–51 Trade Cases ¶ 62,914 (E.D. Mich. Civil No. 8940; Complaint, 1950; Consent Judgment, 1951).

for a patented invention are pushed aside and that section of the Clayton Act which forbids the lease or sale of goods, whether patented or unpatented, upon the condition that the user shall not use the goods of a competitor, is amended or repealed.[92] If the sponsors of this definition did not seek to weaken or cancel the law on this subject, then what was their purpose? In any event, this new definition creates uncertainty about a restrictive practice which in one form or another had been condemned by the Supreme Court, and thus encourages another outbreak of schemes for the revival of the practice.

Criticism

One argument for the patent owner's stipulation of supplies, even at high prices, is that the compensation received by the patentee depends entirely on the amount of benefit which the licensee derives from using the patented article—in other words, the best measure of the benefit is the quantity of supplies used. But the entire remuneration of the patentee does not always come from supplies. Also, the purchaser of the invention often realizes, but too late, that his payments for supplies—usually at prices above the competitive level—are much greater than he anticipated. He then either has to discard the invention or continue to pay high prices for its supplies; ordinarily he continues to pay—as the seller expects.

An alleged advantage of this practice to the licensee is that it distributes the financial cost over a long period and, therefore, makes payments easier and encourages the use of the patented article. The appraisal of this point, like that of the preceding one, depends upon the extent to which the financial reward of the patentee comes from this source and, more importantly, upon the total cost of such supplies in the long run. Moreover, such an advantage to the licensee is offset partly by the inconvenience of the restrictions. Many licensees of the Dick, United, and other companies have indicated their preference for the acquisition of their patented articles without restrictions of any sort.

The restrictive practice enables the patentee to reap high profits from his supplies and hence increases the total financial return from

[92] In this connection it is interesting to note that the Supreme Court in the Times-Picayune case of 1953 stated in a five to four decision that tying clauses are interpreted less strictly if patents are not involved, 345 U.S. 594 (1953).

his invention. The practice means greater prospective gains and therefore stimulates greater efforts to invent and maintain other patents which can likewise be restrictively leased. But this argument overlooks the fact that such restrictions discourage invention and initiative in general with respect to the supplies. For example, one might have invented ink, stencil paper, and the like, even superior to those produced by the Dick company, if there had been an absence of restrictions affecting these commodities.

The privilege to restrict licensees, according to yet another argument, makes possible a superior grade of supplies and repairs, for the manufacturer understands better than others the needs and possibilities of his patented article. The Dick company, for instance, argued that its restriction insured excellent paper, ink, and other supplies, and therefore satisfactory results to the user of the duplicating machine. It declared that the ink sold by others was inferior and ill adapted to the Dick machine. The Peninsular company, by way of contrast, charged that the Eureka company intentionally sold staples especially adapted for its machine and therefore was an infringer. The inconsistency and weakness of this argument is evident.

It should be remembered that patentees exercised this privilege with reference to both patented and unpatented supplies for their machines or other patented articles. It may be true that the owner of a patented chemical process should furnish the delicate materials required, and that the manufacturer of a very intricate article, whether patented or unpatented, should supply the parts and general repairs for it. However, general instructions from the manufacturer as to standards of quality, together with the self-interest of the purchaser of the article, should suffice. As pointed out by the Supreme Court in the International Salt case, an invention is not allergic to salt or other supplies of equal quality provided by anyone else except the patent owner.

Many patent owners maintain that the inventor of a machine or other goods creates the demand for the accessories used with it; that the demand for the accessories did not exist previously; and that the inventor had, therefore, the exclusive right to sell them. This might be true for the first seventeen years after the basic invention.

The mimeograph, salt, shoe machinery, and other cases indicate that the control of supplementary products is not only a problem of monopoly and restraint of trade from the standpoint of the consumer. They also forcefully reveal unfair competition against rivals. The buyer may enter readily into a tying-clause relationship because of the small initial payments—a fact which makes the restriction effective in excluding competitors.

In the final analysis any attempt to tie one commodity to another in selling either of them is fundamentally unfair. It makes the sale of the subsidiary article depend upon the tying clause instead of its own merits. Any article which has not the necessary attributes, such as price and quality, to stimulate its sale and use, does not deserve a contractual or other arrangement which will enforce its sale and use. The very existence of such restrictions suggests that in its absence a competing article of equal or better quality would be offered at the same or at a lower price; necessarily so, for otherwise there would be no occasion for the restriction. Tying clauses attached to the use of a patented article, illustrated by those already described, introduce an artificial element. In short, they lessen competition and prevent the manifestation of economic efficiency.

CHAPTER VII

Forced Validity and Prolonged Monopoly

PREVIOUS CHAPTERS have presented the means by which patent owners have sought to widen the scope of their patents by pooling or cross-licensing them, by consolidating them, and by permitting the manufacture, use, and sale of their patented products under license restrictions governing output, purpose, and price. Specific phases of such restrictions are maintenance of resale prices and control of supplementary products. Incidental statements here and there have revealed efforts to support weak or shallow patents and have indicated efforts to prolong or lengthen the monopoly of the patent owner by acquiring the future patents of inventors, workers, and licensees. This chapter is devoted to the various means of avoiding the judicial appraisal of the validity or depth of patents and means of prolonging a monopoly based on patents.

Extent and Causes of Invalid Patents

A knowledge of the extent and causes of invalid patents is necessary for an understanding of the means used to force the acceptance of patents and to prolong patent monopolies.

Extent. The proportion of patents held invalid has increased within recent years although it has been high for several decades. From 1891 to 1904, 30 per cent of all patents which were adjudicated by the circuit courts of appeal were declared to be invalid, 29 per cent to be valid, and 41 per cent not to have been infringed.[1] During 1941–45, 89 per cent of all patents involved in cases brought up to these courts and the Supreme Court were declared invalid.[2] Are these percentages typical for patents in general? Presumably only patents covering important economic items are infringed and then litigated

[1] 60 Cong., 2 sess., *House Report 2145,* 4.

[2] Daniel G. Cullen, "Patents in Litigation, 1941–45"' *Journal of the Patent Office Society,* Dec. 1946, 903–904.

to determine their validity. Does the patent owner bring action for infringement only for those patents most likely to be upheld by the courts? Do others infringe and defy only those patents most likely to be held invalid? Moreover, if patents on trivial and worthless items were adjudicated, would the percentage of invalidity run higher, lower, or the same?

That federal courts lack confidence in the validity of patents is shown by their unwillingness to protect patents against alleged infringement by granting a preliminary injunction. Patent owners generally are afraid of litigation because of the likelihood of invalidity and also, of course, because of the expense.

A part of the report of a Senate committee describes the situation as follows:

> 1. A considerable portion of some of the patents granted are worthless and void, as conflicting and infringing upon one another, or upon public rights not subject to patent privileges, arising from either a want of due attention to the specification or claim, or from the ignorance of the patentees of the state of the arts and manufactures, and of the inventions made in other countries, and even in our own.
>
> 2. The country becomes flooded with patent monopolies, embarrassing to bona fide patentees, whose rights are thus invaded from all sides; and not less embarrassing to the community generally, in the use of even the most common machinery and long known improvements in the arts and common manufactures of the country.
>
> 3. Out of this interference and collision of patents and privileges, a great number of law suits arise, which are daily increasing in an alarming degree, onerous to the courts, ruinous to the parties, and injurious to society.
>
> 4. It opens the door to frauds, which have already become extensive and serious. . . .

This report was made in 1836 and its purpose was to reveal the defects of the registration system, but part of it is appropriate in 1954 under the examination system.[3]

Causes. As stated in Chapter One, the general prerequisites for a patent are novelty, invention, and utility. If a patent lacks any one of the prerequisites, it is invalid. Additional reasons for invalidity are inadequate disclosure, indefinite claims, and kindred shortcomings.

[3] 24 Cong., 1 sess., *Sen. Doc. 338,* 3.

An alleged invention lacks novelty if known or publicly used in this country or its possessions at any prior time or if it is described in any patent of any nation or in any printed publication anywhere. And, of course, there is no ready means of determining what constitutes a sufficient degree of invention or utility. In appraising an application the Patent Office may not search the records sufficiently or it may misinterpret these concepts.

It is becoming increasingly difficult to ascertain and appraise the records. The number of patents granted by the United States alone had reached 2,700,000 by 1954. Over 99 per cent of them are questionable because their validity has never been passed upon by the courts. Thousands of patents in each important industry reflect an intermingled and interdependent mass of thought, and no one can tell exactly where one ends and the other begins. An inventor once testified that RCA "was in a position to practically invalidate any patents they sought to fight" because "there are very few inventions that are of such a nature that they stand in an act independent of any other work of man."[4]

According to the president of Standard Register Company, getting a good patent requires the following: "The inventor must be thoroughly versed in the prior art; he must know how to properly rough out the specification by writing a thorough and comprehensive description of the matter for the education of his patent attorney; he must know the thing developed, both in structure and in relation to similar developments, that have gone before; he must be able to take the patent attorney's specification and claims and properly correct, interpret, and amplify them, so that the case is fully and properly developed, yet restricted strictly to the thing itself."[5] Can the patent attorney, let alone the inventor, meet these requirements?

This brings up the question of who has the advantage in proceedings before the Patent Office—the applicant or the examiner. The lone inventor and his lawyer too generally know less about the prior art than the examiner. In contrast, the corporation's inventors and technicians and attorneys apparently have the advantage over the examiner. Evidently with this situation in mind, Frank Kottke stated, "The examiner is not a specialist in the sense in which the applicant

[4] Pooling Hearings, Part 4, 3390.
[5] Comp. L. Hearings, 534.

is a specialist, and he is allotted only a few hours to test an application which the latter may have taken months to prepare."[6] Lack of ability or information, no matter on which side it exists, leads to weak or invalid patents.

Accentuating this inequality are three basic weaknesses. The proceedings in the Patent Office are *ex parte*—they are, in effect, primarily in the interest of the inventor. No one is present at every turn to challenge the evidence and other phases of the procedure in appraising an application and granting a patent. Second, several opportunities in the Patent Office are given to the applicant but no time is reserved to review a decision unduly favorable to him. Third, the basic nature of the administrative process itself tends to make the regulatory agency the "captive" of the regulated groups.

After securing and examining all pertinent data, the examiner may approve the issue of a patent which is later declared invalid by the courts. The explanation may lie in a difference of interpretation. A district court recently raised "the question of efficiency of the present patent system which permits the issuance of patents in the number disclosed in this case which, when subjected to the test of litigation, results in invalidating the work of the patent authorities."[7] In this connection it should be added that circuit courts of appeals themselves and also the members of each of them and of the Supreme Court itself occasionally disagree on what constitutes patentability.

Has the Patent Office become less thorough in its search or more liberal in its interpretation of patentability? The evidence indicates that the main change is its increasing inability to perform its duties. As already explained, the problems in determining patentability have grown, and the examiner generally is no match for the patent attorneys and technicians of the corporations which obtain an increasing proportion of the patents.

Is the so-called stricter interpretation of the courts entirely the result of weaker patents? To what extent have judges been influenced by the shift from the lone inventor and his shop to the corporation and its laboratory and its hired inventors and its patents and in some instances its violation of antitrust legislation? Has this shift in the

[6] Frank J. Kottke, *Electrical Technology and the Public Interest,* 39.
[7] *United States* v. *General Electric Co.,* 82 F. Supp. 753, 817 (D.N.J. 1949).

source of inventions made judges more conscious of the rights of the public?

Forced Validity

The clash between applications for patents and between issued patents, the likelihood of defects in either category, and the certainty of expense of controversy either in the Patent Office or in the courts have encouraged sundry schemes for avoiding legal tests in order to maintain patent validity.

Compromises in Interference Proceedings. Applications for defective patents may not be eliminated through interference proceedings even after patent examiners acknowledge conflicts between the applications. In some instances rival applicants settle their differences through compromise rather than by a clash of evidence and a judicial settlement of the issue. In 1934, for example, Hercules and Foote Mineral Company, in interferences with each other over patent applications assigned to them, agreed to settle the issue through counsel and then require the victor to grant an exclusive license to the other under any resultant patent. In effect, they combined and pooled their claims in a single patent.[8] In 1935, Hercules and Du Pont, involved in certain interferences, settled them by filing concessions of priority and by cross-licensing each other under the resultant patents and also under others already issued to them.[9] Standard Oil Development Company and the Chemical Construction Corporation settled an interference by Standard's furnishing a concession of priority to Chemical, and by the two companies cross-licensing each other.[10] Masonite Corporation and Insulite settled interference proceedings by the one conceding priority to certain patent claims of the other and in return receiving an exclusive royalty-free license under all of the other's patents and patent applications relating to hardboard.[11] Vehicular Parking and one of its prospective licensees settled interference proceedings whereby the prospective licensee acquired an exclusive license under certain claims of a patent application. Sometimes rival applicants for patents, illustrated by Hartford-Empire and Owens,

8 Pooling Hearings, Part 4, 3501.
9 *Ibid.*, 3481.
10 *Ibid.*, 3693–94.
11 *United States* v. *Masonite Corp.*, 316 U.S. 265, 273–74 (1942).

shift "claims in such a manner as to obtain the strongest patents possible."[12]

Two examples taken from complaints and consent judgments may be cited at this point. Bearing Distributors and two other manufacturers of tractor cabs ended, in 1948, a three-way interference; one of them was awarded priority of invention, and all of them became licensees.[13] In another case, an applicant for a patent on a telescope grocery cart assigned its application, in 1949, to another inventor and conceded priority to the patent application of the latter in return for a license under any resultant patents at a royalty of twenty-five cents per cart less than any other licensee; as a result the original applicant and two other licensees dominated the business.[14]

These compromises and concessions have the immediate effect of approving questionable claims and therefore of removing each participant as a litigant in determining the validity of the resultant patent. Of course, others who are interested, if any, may subsequently attack a patent's validity.

In extreme cases an applicant may resort to fraud in obtaining a patent. For several years Hartford-Empire sought a patent with which it hoped to dominate the gob-feeding art. The Patent Office refused the broad claims of the application, so Hartford-Empire and Owens sought to change the Office's position. They jointly prepared an article and had it published in a glass trade magazine as the purported work of a labor leader, extolling the virtues of the machine on which the patent was sought. The president of Flint Glass Workers Union signed the article and in return received $8,000. In 1926, patent solicitors directed the attention of the Patent Office to the article as evidence that the Hartford-Empire feeder was a remarkable advance in the art of fashioning glass by machines, but the Patent Office was never told who the real author was. In 1929, the application was allowed and a patent with the broad claims was issued. Later the patent was held invalid for fraud and the attorney involved was disbarred.[15]

12 *United States* v. *Hartford-Empire Co.,* 46 F. Supp. 541, 611 (N.D. Ohio 1942).

13 *United States* v. *Bearing Distributors Co.,* CCH 1952–53 Trade Cases ¶ 67,595 (W.D. Mo. Civil No. 6895; Complaint, 1951; Consent Decree, 1953).

14 *United States* v. *Telescope Carts, Inc.,* CCH 1952–53 Trade Cases ¶ 67,573 (W.D. Mo. Civil No. 6935; Complaint, 1951; Consent Decree, 1953).

15 *Hazel-Atlas Glass Co.* v. *Hartford-Empire Co.,* 322 U.S. 238 (1944).

Once defective patents are issued, however, they are not always weeded out. This is no fault of the courts because they already follow the rule of testing the validity of patents in infringement and other cases which might be disposed of on other grounds. The owner of a defective patent, like the applicant for one, seeks compromise rather than litigation in order to avoid invalidity and expense. Previous chapters suggest how this may be accomplished through pools, consolidations, and license agreements. Owners of rival but questionable patents may cross-license each other or one may purchase the patents of the other and thus lessen or eliminate any occasion for either one to challenge the patent of the other. Additional schemes for avoiding the exposure of defective patents involve the acceptance of a patent's validity by competitors and licensees and inventors, the avoidance of adjudication in case of alleged infringement, and payments for acceptance and proof of validity.

Acceptance of Validity of Patents by Competitors, Licensees, and Inventors. Rival interests under a competitive regime challenge each other's patents which are open to question and the unworthy are eliminated through litigation. However, such an action and result do not occur if all of them in effect are brought into harmony with each other. In that event patents inherently invalid may remain in effect to the date of their expiration. They are permitted to continue long after the exposure they deserve and perhaps never obtain.

In general, the bigger and fewer the rivals in an industry and the more prevalent defective patents, the less litigation to determine their validity. Rivals often have a general understanding or actual agreement to refrain from infringement suits and therefore testing the validity of patents owned or controlled by each other. The likelihood of invalidation of their patents, along with the expense of litigation, encourage rivals to adopt such a plan. Also, friendly settlement of law suits between rivals is often the pretext for agreement in restraint of trade which includes the acceptance of the validity of patents. In addition, there is usually a close relationship between large concerns in the same industry so that each one knows that it can obtain a license under the patents of the other in case a license is necessary. The result is practically the same as the pooling or cross-licensing of patents. This was true of the dominant manufacturers of flat glass and sundry other commodities. Likewise, the cross-license agreement in the gasoline-cracking case, entered into in 1923, provided that "the contracting parties shall acknowledge the validity of all patents that may be issued" to or acquired by these companies prior to January 1, 1937.[16] In other words, the agreement protected from attack patents which might be issued as late as 1937 and which would not expire until 1954. Also, an agreement in 1934 required the Elliott Addressing Machine Company to admit the validity of the Hill patent owned by A. B. Dick Company. Most of these agreements go unquestioned in that small independent rivals ordinarily do not challenge the patents of a large company because of the expense of litigation and uncertainty of outcome.

Such agreements include potential as well as actual competitors. Companies may license each other under patents that may be useful in particular fields; for example, GE granted to RCA rights under GE patents pertaining to radio, and RCA reciprocated by conferring patent rights in the electrical field; each agreed to accept the validity of the other's patents. Therefore, all companies interested in some or all of a group of patents, regardless of the degree of actual competition, may be brought together in one common policy of avoiding their infringement and accepting their validity.

Of the thousands of patents under which GE manufactured or licensed in 1936, only four of them had been passed upon and sustained in the circuit courts of appeals. In some quarters such a record is explained on the ground of "respect for private property." A more realistic explanation is that GE and Westinghouse admitted the validity of each other's patents under which they cross-licensed each other and that 201 electrical and kindred companies as licensees agreed not to question the validity of these patents. In addition, GE, with its numerous patents pertaining to incandescent lamps, granted licenses to manufacture these lamps under all these patents for a general royalty and subject of course to the acceptance of the patents' validity.[17]

In appraising an agreement between National Lead and Du Pont to accept the validity of each other's patents, Judge Simon H. Rifkind said, "The borders of the private domain in titanium are guarded by hundreds of patents, procured without opposition, and maintained without litigation."[18] He described the inevitable consequences. "The

[16] *United States* v. *Standard Oil Co.*, 33 F. 2d 617, 630 (N.D. Ill. 1929).
[17] Pooling Hearings, Part 4, 3610, 3624–26, 3653.

chief spur to private resistance to the grant of patent monopolies has been withdrawn from the parties who, ordinarily, would be chiefly concerned. It is not without significance that, despite the jaundiced eye with which the courts have been examining and rejecting patents in recent years, not a single one of the hundreds of patents here involved has ever been litigated. The result of that is that the newcomer is confronted by a veritable jungle of patent claims through which only the very powerful and stout hearted would venture."[19]

The members of a pool involving patents on variable condensers agreed to accept their validity. In 1949, a district court, in passing upon their practices, said, "Although they contended that the primary purpose of their agreements was to resolve patent litigation, the course of conduct followed by them demonstrates that it was to insulate their patents from outside and inside attacks in order that they might continue a dominating position."[20]

According to a district judge in 1950, "There is slight evidence that pertains to the validity of Alcoa's patents. It appears that none of them has been held valid and infringed by an appellate court of the United States, or held valid and infringed by a district court of the United States after litigation of the issues of law and fact, and not reversed on appeal. . . . Indeed, the last suit for infringement brought by Alcoa was in the year 1936."[21]

Hartford-Empire, Masonite Corporation, United Shoe Machinery, and other companies already described in Chapter Three, have required their licensees to acknowledge the validity of their patents. The licensee of Hartford-Empire had to acknowledge "the validity of the Hartford patents used by him" and each "agent" of Masonite "expressly recognized the validity of Masonite's patents during the life of the agreement."[22]

A patent owner may extend the effectiveness of its patents until the adjudication of the patents by the Supreme Court and therefore may extend the effectiveness into the indefinite future, although lower courts have declared the patents invalid. In 1940, six patents of Car-

18 *United States* v. *National Lead Co.*, 63 F. Supp. 513, 521 (S.D.N.Y. 1945).

19 *Ibid.*, 532.

20 *United States* v. *General Instrument Corp.*, 87 F. Supp. 157, 190 (D.N.J., 1949).

21 *United States* v. *Aluminum Co. of America*, 91 F. Supp. 333, 389 (S.D.N.Y. 1950).

22 46 F. Supp. 541, 615 (N.D. Ohio 1942); and 316 U.S. 265, 280 (1942).

boloy were declared invalid.[23] Nevertheless, this decision did not free Firth-Sterling, a licensee under these patents, because it was bound by contract to accept their validity until the Supreme Court declared otherwise. According to Firth-Sterling, "They (Carboloy) told us verbally that they were probably going to appeal, and the terms of our license were that those patents were not invalid, as far as we were concerned, until a court of last appeal had passed on them. Therefore, until the matter was appealed, we were still bound by our license."[24] Carboloy did not appeal from this decision and at the same time kept their licensees bound to their agreement.

As already noted, avoidance of a test of validity, so far as the patent owner is concerned, generally arises from the prevalence of defective patents. However, a charge of infringement of a patent and the counter charge of invalidity of the patent may be settled by granting a license and accepting validity even if the patent owner thinks his patent is sound. George Baekeland, speaking for his company, said, "It is not because we think the patent is not good, but these lawsuits are time-consuming things and are expensive, and we do not want to be in the patent-defending and the patent-suing business. Where there is a real honest difference of opinion we will grant a license."[25]

A licensee may gladly accept the validity of patents in order to participate in a price-fixing scheme based upon them. In reply to an inquiry as to why Bausch and Lomb continued to pay royalties on worthless patents, Carl Bausch said, "A great many of the patents that we are operating under are very weak, and if we wished to we could break them ourselves and not pay royalties, but good business judgment, I believe, justifies the idea of paying royalties even on weak patents in order to maintain a price-control picture that is helpful." Another licensee pointed out that if the licensor "were forced to bring suit the patents might be knocked out and the whole Ful-Vue license system would topple, and Ful-Vue and Hi Bo products could then be made by anybody at any price, and the nice margin of profit in this business would be lost."[26] The government, as indicated later,

[23] *General Electric Co.* v. *Willey's Carbide Tool Co.*, 33 F. Supp. 969 (E.D. Mich. 1940).

[24] Bone Hearings, Part 4, 1925.

[25] Comp. L. Hearings, 350.

[26] Bone Hearings, Part 7, 3306.

could attack the validity of these patents, but there is no assurance that the attack will be made.

Agreements about validity may involve inventors. The Eastman Kodak Company resorted to contracts with various inventors to insure the validity of its patents. For example, two individuals, Houston and Blair by name, agreed to acknowledge the validity of two patents which Houston assigned to Eastman. Houston was the inventor and Blair a former officer of another camera company. Other companies have resorted to this same practice.

Promises not to question the validity of patents reach their lowest level if they relate to patents the applications for which involve fraud. In one case a company secured a perjured application and exacted promises from the interested sponsors "never to question the validity of any patent that might be issued on that application."[27] A court held the patent void.

Avoidance of Adjudication in Case of Alleged Infringement. One practice is to threaten and even to start a suit for infringement of patents but always to avoid a judicial interpretation of their validity. One phase of this practice is to get rid of alleged infringers of questionable patents by letting them become licensees. The promoter of the monopoly in parking meters, after stating that a certain patent is the only one "that seems to have claims which in any way might be construed as broadly covering a parking meter," added that "these claims are not such, however, as should be put into litigation because I am afraid they might be invalidated by certain prior art. At present they are accorded a *prima facie* validity which could be used to advantage in discouraging competition."[28] This promoter employed both patents and trade-marks in practically forcing Dual into the scheme. "I have felt all along that in order to complete our patent position it is most desirable to bring Dual into the picture. One of the best ways to achieve this result in my mind seems to convince Mr. Magee (Dual) that he is trespassing on certain of our patent and trade-mark rights and then to point out that it is decidedly to his advantage to obtain a license under these various rights."[29]

Complaints and decrees already described in other chapters indi-

[27] *Precision Instrument Mfg. Co.* v. *Automatic Maintenance Mach. Co.*, 324 U.S. 806, 819 (1945).

[28] *United States* v. *Vehicular Parking*, 54 F. Supp. 828, 839 (D. Del. 1944).

[29] *Ibid.*, 832.

cate the avoidance of adjudication in cases of alleged infringement. In one instance, in 1937, the master in chancery in a federal court, after hearing testimony in a suit which challenged the validity of patents on reshaping joint rail bars, held them invalid; thereupon the suit was compromised, legal proceedings ceased, and thus a ruling by a court on their validity was avoided. Furthermore, two companies, which reshaped such bars within the territory granted exclusively to the McKenna Process Company by a patent pool, were advised to discontinue such operations, but they refused. As a result, McKenna and its associates began infringement suits and told all railroads that they had the sole right to re-form rail joint bars, warned them about infringement, and urged them to require these two companies to post infringement bonds. Continued operation by these two companies led to keen price competition. The three members of the patent pool met the situation by granting the two companies licenses in certain regions and all agreed upon uniform prices. Other intruders also received threats of infringement suits, disregarded them, and then accepted licenses upon prescribed terms.[30] American Air Filter Company, another example obtained from complaints and consent decrees, compromised an infringement suit with one company and made it a sublicensee. Another infringement suit was withdrawn and the alleged infringer was granted a license to make a particular type of filter under price and other restrictions.[31]

Examples based only on government complaints may be cited. Bendix Aviation and its affiliates were charged with preventing patent litigation from proceeding to final adjudication when such a test might invalidate or otherwise prejudice their patents.[32] Also, it is alleged, two manufacturers of slip covers for furniture that cross-licensed each other brought at least fourteen patent infringement suits but none was ever litigated, for litigation would have brought about a judicial test of their validity.[33]

30 *United States* v. *Rail Joint Co.*, CCH 1946–47 Trade Cases ¶ 57,469 (N.D. Ill. Civil No. 43–C–1295; Complaint, 1943; Consent Judgments, 1944 and 1946).

31 *United States* v. *American Air Filter Co.*, CCH 1946–47 Trade Cases ¶ 57,492 (W.D. Ky. Civil No. 574; Complaint, 1943; Consent Judgment, 1946).

32 *United States* v. *Bendix Aviation Corp., Westinghouse Air Brake Co., et al.*, CCH 1948–49 Trade Cases ¶ 62,349 (S.D.N.Y. Civil No. 44–284; Complaint, 1947; Consent Judgment *re* Westinghouse, 1948).

33 *United States* v. *Joseph A. Krasnov* (E.D. Pa. Civil No. 11024; Complaint, 1950).

Payments for Acceptance and Proof of Validity. Complaints and consent judgments in antitrust cases point also to questionable means for bringing about the acceptance of certain patents and therefore the continuance of their possible validity.

In 1938, a licensee of a member of the patent pool relating to rail joint bars reshaped such bars for a railroad which refused to pay royalties because of alleged invalidity of the patents in question. After being informed of this dispute, the member-licensor negotiated with the railway. The resultant agreement provided that, if the railroad would pay for re-forming bars according to the rates fixed in the license, the lessor would pay a rebate in the amount of such royalties. This rebate was a secret one and was not given to other railroads.[34]

About 1922, an employee of A. B. Dick Company named Edward W. Hill applied for a patent on a cellulose stencil and assigned his application to his company. Beginning about 1923, similar stencils were manufactured and sold by H. E. Smith and Joseph W. Rowe, and, after the issue of the Hill patent in 1925, Dick sued them for infringement. Smith and Rowe contended that the Hill patent was invalid because of prior art. In 1927, while the patent suits against them were awaiting trial, Dick agreed to pay Smith and Rowe substantial sums of money in installments. Also Dick promised to pay them additional sums unless, by 1931, the Hill patent had been declared invalid by a court of last resort. The obvious purpose of this proviso was to give Smith and Rowe an incentive to establish the validity of this patent. These two men acknowledged the validity of the Hill patent, consented to the entry of a decree enjoining their infringement of it, transferred to Dick certain evidence tending to show anticipation of the Hill invention, and agreed to refrain from rendering assistance to any stencil manufacturer other than Dick. Subsequently in Dick's suits for infringement against two other companies, these men, questioned concerning their agreement with Dick, concealed their pecuniary interest in the adjudication sustaining the validity of the Hill patent and withheld evidence of prior art. The Hill patent was held valid in both actions. The A. B. Dick Company, it is alleged, also acquired and suppressed evidence relating to the

[34] *United States* v. *Rail Joint Co.*, CCH 1946–47 Trade Cases ¶ 57,469 (N.D. Ill. Civil No. 43–C–1295; Complaint, 1943; Consent Judgments, 1944 and 1946).

validity of other key patents employed in promoting and sustaining its monopoly.[35]

Legal Interpretation. In 1892, the Supreme Court, in appraising a license with restrictive provisions, said, "The real question is whether the defendant can stop himself from disputing patents which may be wholly void, or to which the plaintiff may have no shadow of title." In appraising the issue the tribunal declared, "It is as important to the public that competition should not be repressed by worthless patents, as that the patentee of a really valuable invention should be protected in his monopoly."[36]

A district court disapproved the agreement of refiners in the cracking gasoline case to acknowledge the validity of each other's patents. The court declared that "while a patent is presumptively valid, many of them, although duly issued, are invalid for various reasons. The public, in whose interest the patent laws are enacted, is ordinarily protected against the burden of such void patent grants by the action of competitors of the patentee who, prompted by motives of self-preservation, refuse to recognize these void patents, and therefore successfully contest them. The public is thereby relieved of the burden which their existence entails."[37] However, by such agreements these companies "purchased immunity from attack on their patents. Tying the hands and sealing the lips of the only parties who would ordinarily stand suit and contest the validity of the patents," they "attempted to fasten on the public burdens which it was not the purpose of the patent law to impose. Such agreements violate the letter and the spirit of the patent law, and are contrary to public policy."[38] The Supreme Court, to which this case was appealed, did not pass upon this question. It stated, "The propriety of certain blanket acknowledgments of patent validity in the first contract, and in a number of licenses under later contracts, was questioned by the lower court. At its suggestion, these provisions also were formally cancelled by the parties. As the relief here sought is an injunction, and hence relates only to the future, the alleged validity of such provisions has become moot."[39]

[35] *United States* v. *A. B. Dick Co.,* CCH 1948–49 Trade Cases ¶ 62,233 (N.D. Ohio Civil No. 24188; Complaint, 1946; Consent Judgment, 1948).

[36] *Pope Mfg. Co.* v. *Gormully,* 144 U.S. 224, 233, 234 (1892).

[37] *United States* v. *Standard Oil Co.* (*Ind.*), 33 F. 2d 617, 630 (N.D. Ill. 1929).

[38] *Ibid.,* 630.

The validity of a patent may be challenged if a licensee enters into an agreement which would be illegal if unsupported by valid patents. The Sola Electric Company, the owner of a patent for improvements in an electrical transformer, granted a nonexclusive license to the Jefferson Company to manufacture and sell the patented transformers. The contract provided, in effect, that Jefferson's prices should not be less than those of the Sola company and other licensees. This licensee contended that the licensor was not entitled to recover royalties because of the price control provisions and the invalidity of the patent. The issue, as stated by the Supreme Court, was "whether a patent licensee, by virtue of his license agreement, is estopped to challenge a price-fixing clause in the agreement by showing that the patent is invalid, and that the price restriction is accordingly unlawful because not protected by the patent monopoly."[40] It settled the issue and at the same time reversed the lower court by stating that this licensee "may assert the illegality of the price-fixing agreement and may offer any competent evidence to establish its illegality, including proof of the invalidity of the patent."[41] Chief Justice Stone delivered the opinion.

Other opinions of the Supreme Court repeat the stand taken in the Sola Electric case. A decision in 1947 involved a patent license agreement in which the licensee agreed to pay royalties, to sell at prices prescribed by the patent owner, and also not to question the validity of his patent. One licensee ceased to pay royalties on the ground that the patent did not cover its products and also that it was "invalid for want of invention."[42] The patent owner sued to recover royalties. As defined by the Supreme Court, the question was whether the licensee, "in a suit to recover royalties only under a terminated patent license agreement containing price-fixing provisions, can challenge the validity of the patent despite a covenant in the license contract that he would not do so."[43] The court held that the licensee's "obligation to pay royalties and its agreement to sell at prices fixed" by licensor were integrated and must stand or fall together.[44] The

[39] *Standard Oil Co. (Ind.)* v. *United States,* 283 U.S. 163, 181–82 (1931).

[40] *Sola Electric Co.* v. *Jefferson Electric Co.,* 317 U.S. 173 (1942).

[41] *Ibid.,* 177.

[42] *Katzinger Co.* v. *Chicago Metallic Mfg. Co.,* 329 U.S. 394, 397 (1947).

[43] *Ibid.,* 395.

[44] *Ibid.,* 401.

promise to pay royalties was tainted by the price-fixing provision. Further, the fact that the licensee suggested the price-fixing provision did not bar his challenge to the validity of the patent. Finally, a contract, either specific or implied, not to challenge validity of the patent, could not override laws against restraint of trade.[45]

In interpreting its decision in the Sola Electric case the court said the decision "was firmly grounded upon the broad public interest in freeing our competitive economy from the trade restraints which might be imposed by price-fixing agreements stemming from narrow or invalid patents."[46] Also, the court stated, "We need not consider whether under the ruling of *Bement* v. *National Harrow Co.,* 186 U. S. 70, 87–91, these price-fixing provisions would be lawful if the patent were valid. The question here is entirely different."[47] Another decision like the Sola Electric one was handed down on the same day.[48]

In the Gypsum case the government attacked the validity of certain patents. The Sherman Act was violated, it maintained, because invalid patents were used as a basis for restrictive licenses. The trial court held that such attack would constitute a review of action by the Commissioner of Patents which is not authorized by statute. The Supreme Court did not agree with this conclusion. "In an antitrust suit instituted by a licensee against his licensor we have repeatedly held that the licensee may attack the validity of the patent under which he was licensed, because of the public interest in free competition, even though the licensee has agreed in his license not to do so."[49] At this point the court cited the three cases just presented, then went on to say that "in a suit to vindicate the public interest by enjoining violations of the Sherman Act, the United States should have the same opportunity to show that the asserted shield of patentability does not exist."[50] Justice Frankfurter, concurring in the decision, disapproved discussion of this point on the ground that it was not necessary for the disposal of the case.[51]

45 *Ibid.,* 401–402.
46 *Ibid.,* 400.
47 *Ibid.,* 398.
48 *MacGregor* v. *Westinghouse Electric and Mfg. Co.,* 329 U.S. 402 (1947).
49 *United States* v. *United States Gypsum Co.,* 333 U.S. 364, 387 (1948).
50 *Ibid.,* 388.
51 *Ibid.,* 402–404.

Prolonged Monopoly

Previous chapters indicate the means by which patent owners have extended the area of their monopoly. The first part of this chapter reveals the practices of sustaining the effective life or presumptive validity of patents until their expiration—usually seventeen years from the date of issue, though it is true that a monopoly from patents, especially a large one, may be prolonged or continued beyond a seventeen year period. Schemes for this extension of validity are delayed applications, incomplete disclosure, restrictions in terms of the most recent patents, required assignment of inventions by laborers and licensees and former rivals, control of market and laboratories for inventions and of inventors, persistence in unlawful use of patents, and the use of trade-marks.

Delayed Applications. The inventor is allowed six months in which to reply to the official letters of the Patent Office concerning his application. Regardless of the number of answers the inventor makes to the notices of the Patent Office or the number of interferences brought against the application and the time consumed as a result, the prior inventor is entitled to receive a patent on his invention. While the patent is pending, he may begin to manufacture the patented product, or the invention may lie dormant. No one else is likely to develop his invention; however, while the application is in the Patent Office, any one who knows about the invention it covers, unless his information is obtained by improper means, may use it and be free from infringement until the patent is issued. Applications for patents are kept secret in the Patent Office; and though the invention is known to exist, others hesitate to utilize it, especially if much investment is necessary, because they may become infringers when the patent is granted. Moreover, the patent begins to run from date of issue. Delays may extend unduly the monopoly of the invention beyond the statutory period of seventeen years.

As already indicated, either the applicant or others may account for undue delay in the refusal or issue of patents. The delay brought about by the applicant may be attributed to unwarranted claims in the application, the inopportuneness of the invention, and the desire to include subsequent inventions. Two or more of these factors, of course, may be operative at the same time.

The initial claims of a prospective patentee are usually broader than his invention justifies. He may not be acquainted with the art to which it refers but more frequently, the applicant claims as broadly as possible on the theory that it is the Patent Office's job, not his, to narrow down the claim. Lastly, some attorneys, it is said, purposely file one or more unduly broad claims so that the Office will reject them and thus leave a file wrapper record of its having examined the claims carefully.

The time may not be opportune for patenting and introducing an invention, owing to the prevalence of an economic depression or the backwardness of the industry to which it relates. In either case the invention is ahead of its time. The Selden patent, purposely delayed until conditions became favorable for the automobile, was eventually declared invalid. Also, the inopportuneness may relate primarily to an improvement on a basic invention. An improvement cannot be used legally in conjunction with a patented basic invention without the consent of the senior patentee, and therefore a patent on it may not be taken out until about the time that the patent on the basic invention expires. The incentive to do this may be present even though the same company controls both inventions. By graduating the delay in the securement of patents on improvements according to the expiration of basic and other patents, industrial monopolies have been extended over a longer period of time.

An applicant, by "juggling" with an application in the Patent Office several years, may use it as a scoop or dragnet for all similar applications which may be filed in the meantime by other inventors. When a relevant patent is issued to another inventor the applicant immediately becomes busy with the dormant application, "clarifying" the originally loosely drawn specification and claims so as to include the invention which has just been patented. This may not be construed as a violation of Patent Office rules which state that no new matter can be added to a patent application after it is filed, especially if the other inventor is financially unable to fight to the limit for his rights, through appeals to the Board of Examiners and then to a federal court.

Much of the delay in granting patents arises from interference proceedings. They are for the purpose of determining priority of invention between two or more parties claiming substantially the same patentable invention. The interference may occur when different inventors file applications covering the same subject matter. Neither one, as a rule, anticipates a conflict with the other. On the other hand, anyone who files a "dragnet" application expects it to interfere with subsequent inventions; in fact, it is made for that purpose. The preparation, presentation, and consideration of such cases, together with various appeals which may be taken, may require several years. Interferences, it seems, are the chief explanation of the prolonged delays in the Patent Office.

Instances of extreme delay in the granting of patents may be cited. Emile Berliner, who added improved features to the microphone, applied for a patent in 1877 but it was not issued until 1891.[52] In 1920, five patents were issued to Heaney that covered every phase of the starting and lighting system for automobiles. They had been in the Patent Office for fifteen years, and in the meantime other patents had been issued; to play safe, General Motors paid $600,000 for a license.

The applications for the Steimer patents on gob-feeding procedure in making glass containers were in the Patent Office from 1910 to 1916, in interference from 1916 to 1925, before a primary examiner from 1925 to 1927, before the Board of Appeals—which refused a patent—from 1927 to 1929, and then before the Supreme Court of the District of Columbia from 1929 to 1936. This court directed the Patent Office to issue the patent, which was done in 1937. Approximately forty-four years elapsed between application and expiration. The invention covered by this patent was in use at least seventeen years prior to its issue in 1937.[53]

The percentage of applications pending three years or more, whatever the reasons may be, increased from 1.8 per cent in 1890 to 35 per cent in 1930. By 1952, the percentage pending five years or longer had become 3.8 per cent. A study of 143 important patents showing, by decades, the average length of time the applications were pending, shows an increase from nine months in 1890 to nine years six months in 1930. The average period of time of pendency of patent applications increased from one year nine months in 1917 to two years ten

52 FCC Report, 133.

53 TNEC Hearings, Part 2, 439–46, 465, 466.

months in 1929, and to a period from three to three and a half years in 1952.[54] Even as early as 1912, President Taft's commission on economy and efficiency reported, "That there is a considerable percentage of applications which are intentionally delayed by the attorney or the applicant is possible of proof. It may be true that an inventor of small means is anxious to have his application acted on as soon as possible, but even in the case of such an inventor it is not always true that he wishes prompt action, while in the case of applicants who are stronger financially it is alleged the intentional delay is not uncommon.[55]

Whatever the cause, undue postponement in granting patents prolongs the patentee's monopoly and violates the spirit if not the letter of the patent law, particularly if it is sought by the applicant.

Incomplete Disclosure. One way of acquiring and keeping a monopoly is to patent the outline of an invention and yet withhold essential information as to proper temperature, humidity, duration, materials, and the like for its practical application. The scheme is to disclose enough to obtain a patent and withhold enough to make the invention useless to others without the secret know-how. It is possible to withhold such information about hidden processes and intricate apparatus used only in the factory of the patent owner. Of course the first basic patent may be vague because the alleged invention is no more than a nebulous idea. Whatever the cause, the know-how may be more important than what the patent reveals; for example, "The technical knowledge as to glass and surface treatments of the glass is probably much more important than the patented part."[56]

That a license to use an invention and the ability to practice it are two different things is brought out by a 1937 statement of Standard of New Jersey involving an agreement with I. G. Farben: "If for any reason it is not feasible to purchase the catalyst from the I. G., the Standard is free to go ahead with its own methanol catalyst manufacture having a license under the I. G. patents for preparation of methanol catalyst. It is clearly understood, however, that the I. G. cannot give us any of the very special technique and experience which they have developed for the preparation of this catalyst."[57]

54 *Ibid.*, 701–723; and personal correspondence.
55 62 Cong., 3 sess., *House Doc. No. 1110,* 179.
56 TNEC Hearings, Part 2, 661.
57 Bone Hearings, Part 8, 4668.

A patent with incomplete disclosure, though questionable as to validity, helps the owner nevertheless to maintain or continue control in the field which it covers. Witness the difficulty of scientists and technicians in making synthetic rubber in the United States in spite of taking over all the German patents in this field. The monopoly of the enemy continued unduly—and in a time of crisis—because of incomplete disclosure. Research and experiment in this country were necessary in order to find out what the German patents were supposed to reveal. The dominance of the chemical and dye industry in the United States by German companies was perpetuated by taking out patents in this country which did not completely disclose the inventions in question.

In 1935, a former custodian of enemy patents seized during World War I, testified, "Since we took up the patents, more than a million dollars have been spent on finding out how to work them, because always something was left out and always something was covered up."[58] The attempt to utilize the German patents on extracting nitrogen from the air ended in failure because they did not sufficiently describe the processes. The Federal Trade Commission, which administered enemy patents after World War I, reached a similar conclusion about coal tar dyes. According to the commission, "The research work in identifying the processes of manufacturing the various colors and their production in commercial quantities, was not materially aided by the disclosures of the patents."[59] District Judge Arthur J. Tuttle, in holding certain tungsten carbide patents invalid, pointed out, "It was not until the latter part of 1928 or the early part of 1929 that General Electric Company finally put this cemented tungsten carbide material on the market under the name 'Carboloy,' more than six years after it obtained" the patent rights from Germany.[60] He therefore held the patents to be invalid for the further reason that they did not make a sufficient disclosure within the meaning of the patent laws. The German patents on insulin and synthetic quinine were issued several years before these products were developed in Canada and the United States respectively. For some time the possibility of infringing these vague patents discouraged the

58 Pooling Hearings, Part 1, 746.

59 Bone Hearings, Part 3, 1213.

60 *General Electric Co.* v. *Willey's Carbide Tool Co.,* 33 F. Supp. 969, 973 (E.D. Mich. 1940).

research in this country necessary for the creation of these products.

In some instances corporations rely entirely upon secrecy to maintain control of certain inventions on processes and apparatus which cannot be readily examined or understood by employees, while at the same time they take out patents on other inventions which cannot be concealed. Such secrecy, of course, is legal; the patent system is elective, not compulsory. Two American companies, for example, received exclusive licenses from the American Lecithin company under secret processes as well as pooled patents. Each of them agreed to treat these secrets and special information as confidential and prevent their disclosure, and specifically, to refrain from building machinery or equipment used in the manufacture of lecithin for others or from letting others see such machinery. In other instances companies that monopolized or restrained trade have agreed to exchange with each other information concerning "machinery, processes, technology, and inventions." In this arrangement, as in incomplete disclosure, the strategy remains the same—enough patents and secrecy to keep out intruders for many years. Any company, to repeat, which has most or all the important patents and secrets in a field, especially if they pertain to a process, may indefinitely extend its monopoly.

Since 1942, the government in court decisions and consent decrees has ordered licensing under the patents of companies in violation of antitrust legislation. In most instances the government has also required the revelation of know-how to permit competition from beginners. It was realized that compulsory licenses or even cancellation of the patents alone would not be sufficient to terminate the monopolies and restraints in question.

Restrictions in Terms of Most Recent Patents. Control of a product or an industry has been perpetuated by requiring licensees to abide by restrictions until the expiration of the most recent patent and even of future patents.

The agreement between Masonite and its so-called agents provided that it should continue "during the life of that one" of specified patents of Masonite which had "the longest term to run, including any reissues, extensions or improvements thereof," unless it was sooner terminated by either party.[61] The final judgment enjoined the use of patent rights to secure protection beyond the lawful limits of such rights. Other companies have had similar agreements.

In the titanium case the judge said, "The agreements applied to patents not yet issued and to inventions not yet imagined. They applied to commerce beyond the scope of any patents. They extended to a time beyond the duration of any then-existing patent. They embraced acknowledgment of patent validity with respect to patents not yet issued, nor applied for, and concerning inventions not yet conceived. They extended to countries, such as China, where no system of patent monopolies exists."[62] Furthermore, "These patents, through the agreements in which they are enmeshed and the manner in which they have been used, have, in fact, been forged into instruments of domination of an entire industry. The net effect is that a business, originally founded upon patents which have long since expired, is today less accessible to free enterprise than when it was first launched."[63]

A 1929 agreement between Du Pont and ICI provided that each "would, upon request, grant to the other an exclusive license to make, use, and employ within their respective exclusive territories all patented and secret inventions relating to agreement products, 'now or hereinafter during the life of the agreement owned or controlled by the parties.' "[64] In other words, the agreement was based on patents owned hereafter as well as now and thus sought indefinite continuity.

According to a recent complaint and consent decree, a Swedish corporation, in control of gas refrigeration, transferred certain United States, Canadian, and Cuban patent rights to Servel (Electrolux) in return for certain payments. Beginning in 1929, Servel and the Swedish corporation agreed to exchange improvement patents so that each could dominate in its particular territory. This arrangement, it is alleged, was modified in 1946 and 1948 in an effort to avoid prosecution under the Sherman Act.[65]

Required Assignment of Inventions by Laborers, Licensees, and Former Rivals. Another arrangement that tends to perpetuate the industrial monopolies of patent-owning companies already dominant in their fields is the requirement that laborers, licensees, and former

[61] *United States* v. *Masonite Corp.*, 316 U.S. 265, 273 (1942).

[62] *United States* v. *National Lead Co.*, 63 F. Supp. 513, 524 (S.D.N.Y. 1945).

[63] *Ibid.*, 532.

[64] *United States* v. *Imperial Chemical Industries Ltd.*, 100 F. Supp. 504, 539 (S.D. N.Y. 1951).

[65] *United States* v. *Servel, Inc.*, CCH 1953–54 Trade Cases ¶ 67,665 (E.D. Pa. Civil No. 11036; Complaint, 1950; Consent Judgment, 1954).

rivals turn over their improvements to them. Manual laborers are entitled legally to their own inventions and the patents which cover them unless this right is modified by contract. Therefore, manufacturers of importantly patented products usually require their laborers to agree that any inventions which they may conceive during their term of employment shall be assigned to them. In some instances this restriction is extended for a year or so after the termination of such employment. These contracts give command to one source of inventions and thus facilitate an extension of monopoly.

Hartford-Empire's agreements with its licensees specified the following: "(1) Improvements by the licensee on machinery leased and sold shall become the property of the lessor; (2) rights to improvements and inventions covering licensed machinery and processes or methods shall become the exclusive property of the lessor or vendor."[66] According to an analysis of 333 patent contracts vested by the alien property custodian during World War II, 198 contained agreements to license or assign future patents.[67] Similar agreements have been imposed upon rivals whose physical property and patents have been purchased.

In the concrete-block machinery case a district court said, "The intent to restrain trade and the illegality is further augmented and evidenced by the provision . . . granting exclusive rights . . . of all improvements to the existing patents and to inventions still unborn. Such conditions while not illegal, per se, cannot be justified as necessary to the enjoyment of or ancillary to the patent rights conveyed."[68]

Control of Inventors and of Market and Laboratories for Invention. Once a company achieves dominance in an industry through patents, it creates the very condition in which perpetual control of the industry can be maintained. It often provides the principal or only market for the patents of the independent inventor and it becomes the principal or only place of employment for the professional inventor and laborer in the field. The dominating company operates or controls factories where inventions are reduced to practice and where the laboratory and library and guidance are provided along practical lines. In effect, the company commands the main stream of inventive thought

66 *Hartford-Empire Co.* v. *United States,* 323 U.S. 386, 422–23 (1945).

67 EGRW Hearings, Part 4, 604.

68 *United States* v. *Besser Mfg. Co.,* 96 F. Supp. 304, 311 (E.D. Mich. 1951).

and perpetuates its power through patents. Perhaps the most outstanding example of this sort of dominating company is United Shoe Machinery.

Two recent complaints of the government illustrate the control that some companies have over inventors. In 1934, an inventor named Gage, having assigned to Roko Corporation all his patents on rodless or hydraulic oil-well pumps, "recognizes that the monopolies granted by said Gage patents might be seriously impaired if not entirely destroyed in the event that he made inventions or acquired patents relating to rodless pumps" and "agrees that any such" inventions or patents prior to 1944 "shall be the exclusive property of Roko."[69] It is also alleged that IBM systematically pre-empted the services of inventors active in the tabulating field by employing them on a long-term, exclusive-retainer basis.[70]

As basic patents expire improvement patents take their place; witness, for example, the continued control of the incandescent lamp through restrictive licensing under improvement patents. "An improvement patent may, like a patent on a step in a process, have great strategic value. For it may, on expiration of the basic patent, be the key to a whole technology. One who holds it may therefore have a considerable competitive advantage, and one who assigns it and thereby loses negative command of the art may by reason of his assignment have suffered a real competitive handicap. For thereafter he will have to pay toll to the assignee, if he practices the invention."[71]

Persistence in Unlawful Use of Patents. Some patentees have persisted in practices already declared unlawful by the courts. One illustration in detail may be sufficient. In the dry ice decision it was held that the Dry Ice Corporation could not limit the use of its invention to the unpatented materials purchased only from this company. Carbice Corporation, the other litigant, asked the Supreme Court to consider the validity of the patent on this invention because Dry Ice Corporation, after the decision just mentioned, inaugurated a campaign of intimidation against its customers by giving the public press a statement that the true patent monopoly had in no way been lim-

[69] *United States* v. *Kobe, Inc.* (N.D. Okla. Civil No. 2865; Complaint, par. 23, 1951).

[70] *United States* v. *International Business Machines Corp.* (S.D.N.Y. Civil No. 72–344; Complaint, 1952).

[71] *Transparent-Wrap Machine Corp.* v. *Stokes and Smith Co.,* 329 U.S. 637, 642 (1947).

ited by the Supreme Court, that this court had indicated that the proper way to enforce the patent monopoly was by directly suing those who used solid carbon dioxide in the patent combination without a license, and that such suits would be brought immediately.[72]

As previously indicated, various patent owners have continued practices long after they have been outlawed both by statute and judicial interpretation. A general example of this, as already indicated in Chapter Six, is the control of supplementary supplies—stencils, fasteners, caps, etc.—in spite of the Clayton Act of 1914 and repeated condemnation by the Supreme Court beginning in 1917.

Use of Trade-Marks. A monopoly based upon a patent may, in effect, continue after the expiration of the patent. This is especially true of an advertised product of personal use which bears a trade-mark. For seventeen years a trade-mark may be associated with the patented product; the one may suggest the other so invariably that they become practically synonymous in the mind of the consumer. After the expiration of the patent, therefore, the monopoly of the product based on a patent tends to pass over into a monopoly of the same product based on a trade-mark. The same product manufactured by others suggests imitation and inferiority. Aspirin, for example, was advertised as patented and as Bayer's aspirin prior to the expiration of the patent covering the product. The long association of only one trade-mark, that of Bayer and Company, with aspirin enabled this company to continue the domination of the market in this product. Other companies have manufactured aspirin since the expiration of the Bayer patent, but many people who purchase this product still insist upon Bayer's for the reason stated and in spite of its higher price. Of course, if Bayer had been able to exclude others from the use of the word aspirin, as it tried to do, its continued monopoly would have been more complete. Under common law and also under the Lanham Act of 1946 a trade-mark which becomes a common descriptive name or a generic word ceases to be private property.

In an early and somewhat similar case the Singer Company, having enjoyed a monopoly of sewing machines under its various patents and having advertised these machines as made by this company, registered the word "Singer" as its trade-mark for the purpose of keeping a partial control over the business after the patents had expired.

[72] *Carbice Corp.* v. *American Patents Development Corp.*, 283 U.S. 420, 421 (1931).

The Supreme Court disapproved such tactics "to retain in the possession of the company the real fruits of the monopoly when the monopoly had passed away."[73]

Similar strategy is employed in continuing a patent monopoly with the aid of a trivial or substantial improvement. In some instances a manufacturer secures improvement patents, timing their issuance to the date of expiration of the basic patent. He launches a great advertising campaign to "educate" the people about the "improved" product and to differentiate it from its immediate predecessor. He seeks to invest the new with a superior halo and to bring the old into disrepute, and to lead people to purchase the one and to discard the other.

Before the basic patent of the Gillette Safety Razor Company expired in 1919 it launched a campaign in behalf of its new and improved safety razor. The strategy was to have another monopoly take the place of its former one, but Gillette failed because the improvement was trivial. German manufacturers flooded this country with safety razors similar to the old type.

In the early 1900's, General Electric educated the public to use tungsten filament lamps and thus made carbon filament lamps obsolete. Prior to the expiration, in 1929, of the broad patent on the clear lamp, GE obtained patent protection on the frosted lamp. These lamps were alike except for the inside frosting. In spite of the extra expense of frosting, the new type sold for less than the old one. Advertising proclaimed the lower price and alleged improvement. The public demand, as intended, shifted largely from the old to the new after the monopoly cover disappeared from the one and descended upon the other.[74] Ironically enough, the improvement patent, after twelve years of litigation, was held invalid by the Supreme Court after its expiration. More recently a similar campaign has been under way in behalf of fluorescent tubes.

Conclusion

Forcing the validity of existing patents and prolonging a patent monopoly into the future are additional means to defeat both the patent

[73] *Singer Mfg. Co.* v. *June Mfg. Co.,* 163 U.S. 169, 181 (1896).
[74] Pooling Hearings, Part 4, 3836.

and antitrust laws. Keeping thousands of defective patents alive by getting competitors and users of the invention to accept their validity indicates in effect their issue by registration and their maintenance by consent; extending a patent unduly into the future violates in effect the provision for securing exclusive rights to inventors for limited times.

CHAPTER VIII

Suppression of Patents

SUPPRESSION OF PATENTS is the nonuse of meritorious inventions which the patents cover. It usually implies a deliberate decision and a questionable purpose, either of the patent owner or of some one else. It is the shelving of an invention which others would like to manufacture or use if they knew about it, and implies that the patented product or process is as good or even better than any other already in use. The effect of suppression is generally harmful in that it hinders invention or restrains competition or both. Such suppression, like anything else which involves human intentions and speculation about possible results, is difficult to prove or disprove. Nevertheless, it exists according to the evidence available and should be examined in terms of causes, extent, and legality.

Causes of Suppression

The following are the chief causes of the suppression of patents: (1) inadequate finances, (2) incompetence of patent owners, (3) illegality of an unauthorized use of supplementary inventions, (4) expense of replacing old equipment with new inventions, (5) monopoly of patents in an industry, and (6) delay in development of invention. Two or more of these causes may be operative at the same time.

Inadequate Finances. Defending an invention means expensive litigation. Finding and removing its defects calls for many experiments. P. T. Farnsworth and his backers spent a million dollars in developing electronic television over a period of thirteen years.[1] Also, marketing an invention requires sales promotion, demonstration, and kindred efforts. People are dubious of any new product, regardless of its intrinsic qualities; even the army and navy at first refused to accept and develop the airplane as a weapon of war. The independent in-

[1] TNEC Hearings, Part 3, 984.

ventor himself is proverbially poverty-stricken, and he experiences great difficulty in securing financial aid from others.

Incompetence of Patent Owner. The incompetence of the inventor requires little comment. The inventor is seldom a man of business ability. He may dream of the far-reaching effects of his invention, of the wealth which will be his, and of the good fortune of anyone who may share the honor of developing it. His dream is usually broken by stern realities. He may become indifferent to what seems an unappreciative world and, therefore, let his invention lie idle. Moreover, some inventors are ignorant even of the possible channels for marketing patents.

Illegality of Unauthorized Use of Supplementary Inventions. The consummation of every basic and pioneer invention of importance is followed by diverse improvements; but if these cannot be legally employed in conjunction with the former—as often happens when the owners of the improvement patents cannot secure licenses from the owner of the basic patent and cannot or will not sell or lease them to him—inventions must be suppressed or the law defied. If a broad application is delayed in the Patent Office longer than narrow improvements, the broad application will control the improvements even after the patents upon them have expired. Patent pools and consolidations in some instances refuse to grant licenses to outsiders with improvements. On the other hand, obstructive patents covering improvements may in extreme situations necessitate the abandonment of the basic invention. This second situation may be the result of a deliberate policy known as "fencing in"—a policy of taking out patents along the various lines of developing the basic invention and thus closing the best if not all roads leading to its practical or commercial application. The owner, hedged in at every turn, may let his invention lie idle rather than let others take it upon their terms.

Expense of Replacing Old Equipment. Every old enterprise tends to keep the same equipment to avoid the expense of replacement. Harry Jerome indicated that modern capitalism encourages suppression: "Technical progress far outruns actual practice. This margin of nonuse is in part due to nonpecuniary factors but the major explanation is simply that, on the whole, industry must be conducted with profits as the immediate goal; hence the first and major consideration in any choice of method is not merely, will it do the work, but also will it pay?"[2] As stated by patent counsel of General Motors: "It, of course, might happen that an improvement amounting to invention is made, especially in respect to a machine or method of manufacture which is very good, but the use thereof would require the scrapping or discarding of a large quantity of machinery, the replacing of which would be so costly as to make it unprofitable to use the invention."[3]

If companies with the old method or product provide the only market for certain patents and they refuse to buy them, suppression is inevitable of course until this situation changes. J. P. Morgan, after a demonstration of the telephone, said in behalf of the officials and engineers of Western Union, "Mr. Bell, after careful consideration of your invention, while it is a very interesting novelty, we have come to the conclusion that it has no commercial possibilities."[4] The balloon tire was offered to all tire companies in 1914, but was not adopted until 1926. General Electric and also other companies, already successfully engaged in manufacturing and selling their own washing machines, turned down the opportunity of buying patents upon a machine for rinsing and drying as well as washing clothes because the idea was impractical. Bendix, without any vested interest in this field, accepted the challenge.

The new and perhaps small enterprise is more aggressive than the large corporation in developing improvements, alternate methods, and even basic inventions. The one has little or nothing tied up in an investment threatened with obsolescence while the other has much. One inventor and manufacturer of oil-field equipment testified, "The bulk of patents are developed by small outfits, not the larger ones."[5] Another stated, "Now, these inventors would have no place to go at all unless they came to the small shops for development of their ideas."[6]

Change in technology may mean obsolescence in labor as well as in machines, and therefore workers have opposed inventions that threaten their jobs; for example, in Chicago, "Organized skilled work-

[2] Harry Jerome, *Mechanization in Industry,* 331.
[3] TNEC Hearings, Part 2, 692.
[4] Pooling Hearings, Part 1, 685.
[5] Comp. L. Hearings, 166.
[6] *Ibid.,* 95.

ers in the building trades have slowed down the introduction of processes that threaten to endanger their health, destroy their skill, lower their wages, and cause technological unemployment."[7]

Monopoly of Patents in an Industry. Suppression of many patents is incidental to acquiring and maintaining and extending a monopoly of patents in a particular field. These patents may be obtained from hired and independent inventors, workers, and licensees. Some of the inventions which are never produced may be as good as the ones manufactured and sold. Also, such a patent-owning monopoly indirectly may force an inventor to suppress his patent simply by refusing to buy it. In addition, thousands of patents have been taken out in the United States by foreigners solely for the purpose of reserving this country as a market for their inventions manufactured abroad. From the standpoint of manufacture in this country such patents are certainly suppressed.

A district court, in appraising the policies of Hartford-Empire, declared, "There are but two objectives that a corporation could have in acquiring an aggregation of all the patents relating to a particular industry—(a) to gain control of the industry and thereby eliminate all competition, or (b) to suppress those patents which it senses would be detrimental to its ultimate aims and desires."[8] This statement may apply also to patent pools. Other advantages of course from such concentration of patents is less likelihood of infringement suits and tests of validity.

One purpose in getting and keeping control of an industry is to prevent the converse of Gresham's law in technology—to keep the new and better from driving out the old and inferior. An effective way of doing this is the control of alternate products and methods and even of substitutes by means of patents. A company with such protection from competition is not forced to develop a better product. In fact, it is apt to shelve or refuse new proposals "to save the cost of retooling for production and the merchandising effort necessary to present a new product to the public."[9]

Delay in Development of Invention. If the development of meritorious inventions is unduly delayed for the foregoing reasons, some

[7] R. E. Montgomery, *Industrial Relations in the Chicago Building Trades,* 168; NRC Report, 59.

[8] *United States* v. *Hartford-Empire Co.,* 46 F. Supp. 541, 618 (N.D. Ohio 1942).

[9] MP Hearings, Part 2-A, 349.

of the undeveloped patents may be surpassed by better and more recent ones. In other words, inventions may acquire a comparative inferiority in time and become unworthy of development. The nonuse of patents for so long a time that they become obsolescent amounts to suppression. This situation applies especially to mechanical improvements with many variations. "It has been my experience in the electrical industry," said the president of an independent electrical company, "that the life of improvements is very limited. As a matter of fact, I have had improvements that I have made myself absolutely nullified and eliminated by further improvements made by myself or made by others before the patent on the initial improvement is even issued from the Patent Office. In other words, an improvement may not have lasted more than a year or two years."[10] The machine upon which the improvements are based may itself become obsolete and thus destroy their utility. Improvements, for example, upon the crank for automobiles became worthless upon the adoption of the self-starter. Even basic inventions may be swept aside by the moving stream of science and technology. In a progressive competitive industry, such as petroleum refining, machinery and processes generally become obsolete within seventeen years. Think of the many important inventions which may become outmoded because of the application of atomic energy. The extent to which inventions generally become inferior and even worthless year by year is shown in other countries by the increasing extent to which patents lapse from date of issue to expiration because they are no longer worth enough to justify the payment of yearly taxes upon them.

Some argue that if meritorious improvements were suppressed, new factories would arise upon the expiration of the patents on them. But these arguments do not take into consideration that within a few years inventions generally lose much of their utility and some of them become obsolete. Enterprises do not start new plants if forced to begin several years, not to mention seventeen, behind a moving front of technical progress.

Extent of Suppression

The extent of suppression cannot be readily ascertained. One difficulty is in deciding exactly what suppression is and in finding a proper

[10] Comp L. Hearings, 495.

definition of the practice. Operating concerns, especially those with hundreds of patents, usually interpret the practice of suppression in terms of profit only; more specifically, would it pay to develop the new invention and discard the old one already in use? From the standpoint of profit, one patent is superior and the other is inferior. But the contention that only inferior patents are suppressed is weakened by the fact that owners often desire to bring suits for alleged infringements of their patents on suppressed inventions as well as on inventions used. Presumably such suits are not always just for harassment, and the insincerity of the contention that only inferior patents are suppressed is revealed in the owners' opposition to the proposal that their suppression be a sufficient cause for infringement on the part of others. Actually, suppression should be judged not in the light of profits it makes possible for the patent owners but in the light of what effect the suppression has on the industrial arts. As a rule, we refuse to accept the opinion of vested interests as a criterion in any field. The sound and reasonable test to determine whether or not real suppression is taking place is to discover if there is a willingness on the part of others to develop the unused inventions and thus advance the technical arts. If others are willing to use the inventions, then the failure to use them by the patent owners is suppression.

Another difficulty in gauging the extent of suppression, however, is that the quality of an invention is comparative and therefore changes as subsequent inventions occur. As already indicated, nearly all worthwhile inventions become obsolete in course of time, and some of them, especially minor improvements, become obsolete even prior to the expiration of their patents The refusal to develop or use such inventions may be described as suppression only until the inventions are surpassed by others; after they have been surpassed, their not being used is only the result of their inferiority. According to a manufacturer of automobile wheels, his industry "changes so rapidly and competition is so intense and so fierce in the industry that nothing that he patents today is going to continue in the form in which he patents it."[11]

Instances of suppressed patents have been mentioned in explaining the causes of the practice and in describing patent pools, consolidations, and license agreements. Additional examples, in which both American and foreign companies play a part, will be presented in greater detail.

Patents Granted to Americans. Although a complete investigation of the extent of suppressed patents has never been made, considerable evidence may be found in government reports, congressional hearings, analysis of patents of large corporations, court decisions, consent decrees, and complaints of the Department of Justice.

a. *Government Reports and Hearings.* American Tobacco Company acquired a patent for a tobacco-stemming machine by purchasing a majority of the stock of the owning company; this machine was not manufactured, but the control of the patent by the owning company prevented its development and use by competitors.[12] It was General Mitchell's opinion that members of the aircraft pool would not put a great improvement into production unless forced to do so "because they would have to change their old jigs, fixtures, and everything else."[13] In radio, various purchases and license agreements enable RCA to bring together "innumerable competing patents."[14] Two-thirds of the patents of International Harvester were not in current use—"In a relatively few cases the patents were taken out to record priority in a new conception which might be usable but for which no practicable commercial use has yet been found or developed."[15] According to Ingersoll-Rand, "Our patent department files an application on practically everything which is developed at the shops and appears patentable. The engineers embody the inventions in the product if they consider it advantageous."[16] Granting that most of the inventions referred to in these statements are inferior or even worthless, one may safely assume that others are not feasible solely because they would endanger existing investments. Would these companies permit prospective rivals to develop these dormant inventions? That is the test.

According to a memorandum of Hartford-Empire, "In the early days of 1912 and 1913, Mr. Peiler developed definite ideas as to both the paddle feeder and the plunger feeder. For various reasons, those then in charge deemed it wisest to proceed with the commercial development of the paddle feeder. Mr. Peiler had demonstrated with

11 TNEC Hearings, Part 3, 1049.
12 *Tobacco Industry,* Corporation Bureau *Report* (1909), Part 1, 84.
13 Pooling Hearings, Part 1, 838.
14 *Ibid.,* Part 4, 3391.
15 *Ibid.,* 3412.
16 *Ibid.,* 3453.

tests the feasibility of both paddle and plunger. It was a case in which, as it has now turned out, the poorer of two alternatives was selected for the commercial development. The better (plunger feeder) principle was pushed aside for the time by the paddle feeder development. The plunger feeder was not actively worked upon again until 1916 and later."[17]

American Telephone, after obtaining thousands of patents upon various alternate methods of accomplishing specific results, used some of them and let the others lie dormant. Once a selection was developed, standardized, produced, and installed on a large scale, it acquired subsequent inventions that threatened its position, not to develop and use, but to protect its position and salvage its investment in old equipment. To maintain that only the used inventions were worthwhile assumes an omniscience and disinterestedness contrary to fact.

Illustrations of the inertia to large-scale change of standardized telephone equipment are found in the delayed and graduated introduction of the combined handset and the dial system. The combined handset has been in use in Europe ever since the beginning of telephony. "This form of subscriber station apparatus was made commercially available in America by an independent manufacturer in 1925. Shortly thereafter, similar sets were put on the market by other independents. The American company began the introduction of the combined handset to Bell System subscribers in 1937, or twelve years after this form of set was available on the independent market. No systematic replacement of the old type handset with separate bell box, most of which have been installed since 1928, is contemplated, and it is anticipated by the company that the old form of instrument will continue to serve Bell subscribers until removed in the course of the normal conduct of business."[18] It should be added that one phase of this "normal conduct," which permits a wearing out of the old equipment, is an extra charge to the subscriber for the new set. The same inertia to change exists in other parts of the telephone industry: "The introduction of improved automatic types of toll equipment into the Bell System plant will either be delayed or be rendered more costly by the existence of large recent investments in the older forms of manual toll boards."[19] These illustrations may be attributed to the

[17] TNEC Hearings, Part 2, 776.
[18] FCC Report, 259.
[19] *Ibid.*, 262.

inadequate and ineffective regulation of public utilities as well as to patents.

In 1935, this company owned 46 per cent of the patents classified as "telephone" by the Patent Office and 62 per cent were unused. Corresponding percentages for its telegraph patents were 45 and 65; for radio, 15 and 58; and for sound recordings and reproducing, 17 and 50.[20] There is nothing to indicate any important change in these figures in more recent years. Does any one really believe that all these unused inventions were inferior? Would American Telephone have let independent telephone companies develop and use them?

By acquiring the inventions of its own employees and also of others Standard Oil of New Jersey controlled pour depressants. According to correspondence between employees of this company, "It seems to me that our best hope for protecting our patent structure under the existing circumstances is intensive research work to uncover all the possible pour depressants and to literally 'flood' the Patent Office with our applications. There is no such thing as filing too many applications. There is, however, great danger in filing too few applications."[21] In 1932, Mid-Continent Petroleum obtained patents on such depressants, and in the same year Standard negotiated a contract for an exclusive license under Mid-Continent's patents, and in return agreed to supply requirements which it could provide in this field at a discount of 33⅓ per cent from the current sales prices and to give Mid-Continent 25 per cent of the net profits from any sales of products covered by the patents. According to correspondence in 1937, however, Standard had never done anything with the patents it had licensed.[22]

As Standard's handling of Mid-Continent patents indicates, acquiring and suppressing patents on alternate inventions to prevent competition may intensify the effort of others to conceive and develop another rival invention, and this in turn may necessitate another round of purchase and suppression. An official of Standard of New Jersey described this situation as follows:

One question which immediately presents itself is whether, if we buy the Fischer, Pott and Uhde processes, we shall not find that in another year

20 *Ibid.*, 234–35.
21 Bone Hearings, Part 4, 1759.
22 *Ibid.*, 1760, 1812.

or two we have up the same question with respect to other newly developed competitive processes. The probability is that this will be true to some extent, and that any restriction or repression which we exercise over the normal growth of these synthetic processes will result in encouraging the development of competitive processes. All that can be reasonably expected is that the importance of future competitive processes will be on a descending scale, as is usually the way in the development of any new art.[23]

Improvement and alternate patents are obtained to acquire and perpetuate a monopoly. As technology changes and patents expire, new inventions and fresh patents are necessary to provide a continuous protective cover. Throughout the procedure, worthwhile inventions are suppressed until outmoded.

General Electric, the chief manufacturer of incandescent lamps, co-operated with the electric utilities in delaying the introduction of fluorescent lamps. According to a report prepared for the Edison Electric Institute, "When the economics do change and fluorescent lighting becomes cheaper for the customer, utilities will be getting a much smaller share of the consumer's lighting dollar and only about one-half the revenue for the same amount of light. Therefore, it does not seem safe to adopt merely a policy of watchful waiting. Immediate action is needed to meet the situation."[24] However, GE could not afford to continue such a policy after Hygrade Sylvania threatened to become the leader in making and selling this new kind of lighting equipment.[25]

Opposition to fluorescent lights in order to maintain sales has its counterpart in the reduced quality of flashlight lamps to increase sales. A memorandum of 1932 from the files of GE reveals the following:

Two or three years ago we proposed a reduction in the life of flashlight lamps from the old basis on which one lamp was supposed to outlast three batteries to a point where the life of the lamp and the life of the battery under service conditions would be approximately equal. Some time ago the battery manufacturers went part way with us on this and accepted lamps of two battery lives instead of three. This has worked out very satisfactorily.

We have been continuing our studies and efforts to bring about the use of one-battery-life lamps. I think you will be interested in the attached analysis . . . covering the various points involved in going to the one-battery-

23 *Ibid.*, Part 7, 3356.
24 Bone Hearings, Part 9, 4759.
25 *Ibid.*, 4759–4800.

life basis. If this were done, we estimate that it would result in increasing our flashlight business approximately 60 per cent. We can see no logical reason either from our standpoint or that of the battery manufacturer why such a change should not be made at this time.[26]

Comparative and partial rather than absolute suppression exists when a domestic corporation restricts home production in order to stabilize price but permits foreign licensees to produce without restriction in order to receive maximum royalties. And in pursuing such a policy, a domestic corporation of course may unwittingly conform to the scheme of another nation. The New Jersey Zinc Company used its patents on processes for producing high-grade zinc to restrict the supply of such zinc in this country and therefore to stabilize prices, while at the same time New Jersey Zinc encouraged unlimited production under these same patents in foreign countries, including Germany, in order to secure the greatest revenue from royalties. An agreement between New Jersey and each foreign licensee not to invade each other's home market made it possible for New Jersey's two contrasting policies to operate at the same time without conflict. The restriction here, resulting in a shortage of zinc and in the capacity for refining it, handicapped this country at a crucial time. New Jersey Zinc was led, as if by a Hitler hand, in adapting its policy to a strategy which broke into war in 1939.[27] In discussing Anaconda's request for a license, New Jersey Zinc stated in a memorandum that its principal interest is "to avoid the production of additional high purity zinc."[28] That the unlimited production of high-grade zinc by its licensees in Germany helped to build the war machine of that country is implied in the boast of Deutscher Volkswirt, a business publication: "The transition from copper and brass to zinc, in particular in small die castings and fittings products, has favored Germany; she is now the greatest producer of zinc in the world after Poland has been destroyed."[29]

Another instance of high price, low output, and partial suppression involves Carboloy and tungsten carbide. In a crucial time, in 1939, there was a high-priced scarcity of tungsten in the United States in

[26] *Ibid.,* Part 1, 630.
[27] *Ibid.,* Part 3, 1531–55.
[28] *Ibid.,* 1532.
[29] *Ibid.,* 1551.

sharp contrast to the low-priced abundance in Germany. However, production increased rapidly from 1939 to 1941 in response to lower price and greater sales.[30]

According to a report by United Nations on restrictive business practices, "The main basis for assuming that the suppression of technology does take place in the international sphere are the large number of national statutes providing for remedial action where a local patent is not exploited, and the fact that these statutes were originally inspired to a large extent by the failure of foreign holders to exploit such local patents."[31]

b. *Analysis of Patents of Large Corporations.* An examination of the patents of a corporation may reveal the practice of buying up alternate inventions, using some of them and suppressing the remainder. These alternates may be improvements or basic innovations; whether they are superior or inferior, comparatively speaking, may never be known because they are not put into practice. Vested interest in existing equipment and dominance in the field are most conducive to this practice.

The A. B. Dick Company acquired patented inventions relating to sheet-feeding apparatus used in connection with a duplicating machine. The inventions could not be conjointly used in a single sheet-feeding apparatus nor was one the outgrowth or development of another. They represented different lines of improvement of different inventions: some of them related to the feeding of sheets to a duplicating machine from the top of the pile, and others, from the bottom. Those related to the feeding of sheets from the top were put to use but not the others. Dick also had patents pertaining to stencil sheets. These could not be used together in the preparation of stencil paper; they represented different lines of invention, all independent of one another. They covered five methods of forming characters in the stencil sheet—abrasion, pressure, heat, adhesion, and extraction. One method did not become obsolete and then another appear; improvement or evolution of the one did not cease before that of another began. All developed simultaneously, but without being interrelated. Only two methods, pressure and abrasion, have been employed com-

30 *Ibid.,* Part 4, 1873.

31 *Restrictive Business Practices,* United Nations Economic and Social Council *Report,* 13–14.

mercially. The acquisition of the other three methods indicates an intention to monopolize all patents relating to the preparation of stencil paper. Such a policy of getting all the patents in a field generally results in suppression.

Throughout its history the Dick company, according to a government complaint already presented in previous chapters, has purchased patents and obtained exclusive rights under them. Contracts in 1932 as well as in 1899 and 1913 provided that the Klabers would transfer to Dick for use in the United States all its future patents and inventions useful in the stencil-duplicating industry. In 1928 and 1931, as well as 1903, Gestetner Ltd. made a similar agreement. In 1931, Dexter Ltd. granted Dick an exclusive license under its existing and future tissue patents for use in stencil duplication. It is likely that these opportunistic acquisitions of patents covered diverse and competing products and methods as did those of an earlier period and that some of them were suppressed to protect Dick's investment in existing equipment and to preserve its monopoly.

c. *Court Decisions*. A New York court, in 1897, said that National Harrow controlled eighty-five patents on spring tooth harrows, some of them unused.[32] A circuit court, in 1906, ascertained that Indiana Manufacturing had acquired over one hundred patents relating to straw stackers which could not be used conjointly, and hence many of them had been suppressed.[33] In the so-called lock case a circuit court announced, in 1909, that "it is a fact, familiar in commercial history, that patent rights have a commercial value for purposes of extinction; that many patents are purchased in order to prevent the competition of new inventions and of new machines with old machines already installed."[34] The button-fastener and paper bag cases also presented evidence of the suppression of patents.[35] A circuit court of appeals stated, in 1932, "It is common knowledge that meritorious inventions are very often relegated to oblivion through the machinations of well-entrenched and powerful competitors."[36]

[32] 47 N.Y.S. 462 (1897).

[33] *Indiana Mfg. Co.* v. *J. I. Case Machine Co.,* 148 Fed. 21, 24–25 (C.C.E.D. Wis. 1906).

[34] *Blount Mfg. Co.* v. *Yale and Towne Mfg. Co.,* 166 Fed. 555, 560 (C.C. Mass. 1909).

[35] 77 Fed. 288 (6th Cir. 1896); 150 Fed. 741 (1st Cir. 1906); and 210 U.S. 405 (1908).

[36] *Keystone Steel and Wire Co.* v. *Commissioner of Internal Revenue,* 62 Fed. 2d 458, 460 (7th Cir. 1932).

In discussing a price-fixing arrangement among manufacturers of synthetic boards, the Supreme Court, in 1942, after observing "some of whom had competing patents," stated, "This kind of marketing device thus, actually or potentially, throttles or suppresses competing and noninfringing products."[37]

According to a district court, Hartford-Empire "acquired from others a total of 317 patents, and many of these were competing patents. In many instances, upon acquisition, the patents were shelved and the manufacture of the patented article discontinued. . . . In the case at bar, the patents acquired were suppressed for the purposes of attaining control over the industry and for the purpose of eliminating competition."[38] This company also suppressed many patents upon its own inventions and forced others to suppress theirs. The purposes in taking out patents as described by its president and repeated by a lower court were as follows:

> (A) To cover the actual machines which we are putting out and prevent duplication of them;
>
> (B) To block the development of machines which might be constructed by others for the same purpose as our machines, using alternative means;
>
> (C) To secure patents on possible improvement of competing machines, so as to 'fence in' those and prevent their reaching an improved stage.[39]

The Supreme Court, upon appeal of this case, referred to the practice of certain glass companies in applying for patents to "block off" or "fence in" competing inventions. In their co-operative effort "to obtain dominance in the field of patented glassmaking machinery, many patents were applied for to prevent others from obtaining patents on improvements which might, to some extent, limit the return in the way of royalty on original or fundamental inventions."[40] According to this same tribunal, United States Gypsum acquired a "number of other patents relating to the process of making closed-edge board."[41] Some of them covered alternate ways of closing the edge of the board which were not used. In 1948, the Supreme Court said, "Starting in 1943, an impasse developed in the affairs of American Scophony. It

[37] *United States* v. *Masonite Corp.*, 316 U.S. 265, 280, 281 (1942).

[38] *United States* v. *Hartford-Empire Co.*, 46 F. Supp. 541, 618 (N.D. Ohio 1942).

[39] *Ibid.*, 611.

[40] *Hartford-Empire Co.* v. *United States*, 323 U.S. 386, 431–32 (1945).

[41] *United States* v. *United States Gypsum Co.*, 333 U.S. 364, 371 (1948).

stemmed from the failure and unwillingness" of General Precision Equipment Corporation and Television Productions, both affiliated with the motion picture industry, "to exploit the Scophony inventions themselves and their refusal to modify the agreements to permit the licensing of other American firms under the inventions."[42]

According to a district court in 1949, GE bought many patents and patent rights from others, especially from foreign lamp manufacturers, which were never used but which did further a policy of domination.[43] Another district court said that the rate at which United Shoe Machinery "has improved old machine types and has introduced new machine types does not create a formidable record."[44] This court felt that "it cannot be said to be entirely without significance that the median age of currently commercial models is twenty-eight years, and the important machine types (in terms of revenue) are even older."[45] The decision of the court was that "on the whole, the rate of improvement of old machine types does not justify special encomia."[46]

Additional proof of suppression is the suits brought for infringement of patents on dormant inventions. Some companies sue—to prevent infringement or to harass small competitors, or both—on patents they themselves do not use. The actual infringement of dormant patents, in some instances, indicates that they actually are worthwhile and have been suppressed for that reason.

d. *Consent Decrees.* Decrees involving companies already described and discussed indicate the suppression of patents. The everlasting match, invented about 1932, was never manufactured and sold in the United States by any member of the match cartel; Diamond Match, as already indicated in Chapter Five, believed that it was "a distinct danger to the American match industry" although experience showed that its odor prevented its general acceptance by the public.

The four manufacturers of high-tension cable and accessories who cross-licensed each other under their various patents discouraged the commercial exploitation of patents, inventions, and technology of

[42] *United States* v. *Scophony Corp.,* 333 U.S. 795, 800 (1948).

[43] *United States* v. *General Electric Co.,* 82 F. Supp. 753, 806, 847, 857 (D.N.J. 1949).

[44] *United States* v. *United Shoe Machinery Corp.,* 110 F. Supp 295, 331, (D. Mass. 1953).

[45] *Ibid.,* 331.

[46] *Ibid.,* 331.

rival products in actual or potential competition with their own. One of four bought the patents of a Swiss company on a superior and more economical type of gas-filled cable. However, according to the government complaint, this company refrained from commercial exploitation of the cable and refused to let others make and sell it. A consent judgment in 1948 ordered all four companies to grant licenses under their patents and enjoined them from maintaining any agreement to "suppress commercial development or exploitation of patents, patent applications, inventions, or technological information relating to fluid-filled cable or accessories."[47]

e. *Complaints.* Various complaints previously presented, allege that patents are suppressed. Bendix Aviation, in preserving its allocation of business with Wagner Electric, acquired and suppressed a rival set of patents on a hydraulic-braking system; it obtained patents to, fence in braking systems developed and manufactured by others. Du Pont acquired patents on cellophane, not to use the inventions which they covered, but to obtain a monopoly of patents in the field. Also, several companies pooled competing methods for producing elastic-top hosiery and one of them was suppressed. The Roko Corporation paid royalties on an exclusive license under certain patents on a hydraulic oil-well pump though it never had any intention to employ the inventions in question; these and other patents acquired covered alternate types of pump construction.

Patents Granted to Foreigners. The extent and effect of patent suppression are most readily appreciated when one considers the "non-working" of United States patents by citizens of other countries. Many foreigners have obtained patents from the United States, not to utilize them in this country, but to "work" them elsewhere, usually in their own country, from which they have also secured patents on the same subject matter. In this manner, foreigners have reserved the United States as the exclusive market for their goods—a territorial monopoly, as it were, defined by national boundaries.

The extent to which United States patents may be suppressed by foreigners is suggested by the large number of patents, both absolutely and relatively, which are granted to them by this country; for

[47] *United States* v. *General Cable Corp.*, CCH 1948–49 Trade Cases ¶ 62,300 (S.D. N.Y. Civil No. 40–76; Complaint, 1947; Consent Judgment, 1948).

example, even from 1910 to 1915 the number was 21,075 or 11.4 per cent of the total. During 1910–15, 1930, 1940, and 1950 Germany received 33, 34, 42, and 0.1 per cent respectively of all patents granted to foreigners. The extreme drop in 1950 reflects, of course, the dislocations caused by World War II, and does not mean that the Germans have discontinued research. From 1946 to 1950 there was doubt about whether they, and also the Japanese, could establish priority in their inventions from filings in their own patent offices—a doubt removed by Public Law 619 of the Eighty-third Congress; the foreign exchange restriction at first precluded and later made difficult their procurement of patents. Since these and other obstacles have been overcome entirely or partly since 1950, Germany's proportion should markedly rise although loss of scientists and technicians resulting from the sovietization of East Germany and from emigration will make it difficult for it to regain its former standing. It is likely, however, that some patents on German inventions have been taken out through nearby nations. Worthy of note is that the number of patents going to Switzerland increased from 4.7 per cent of the total granted to all foreigners, in 1935, to 10.4 and 9.7 in 1945 and 1950 respectively. The actual numbers for these three years were 284, 219, and 429.

The relatively large number of patents granted to foreigners by the United States may be ascribed partly to the cheapness and convenience of acquiring and maintaining patents in this country. A second reason, and a more important one especially since World War I, is that this country affords the best market for making sales or licensing patents and thereby earning dollars.

a. *Before World War I.* Most of the patents granted to foreigners, to Germans especially, related only to a few subjects, the chemical and dye industries in particular. This fact, together with the suppression of such patents in the United States, largely accounts for the lack of development of these industries in this country before World War I. Our supply of chemicals and dyes came mostly from Germany —from the country which owned most of the United States patents on these products. Our industry of dyestuffs and medicines consisted largely of small assembling plants operating on German intermediates and therefore at the mercy of German producers. At the beginning of 1914, German concerns supplied nine-tenths of the dyes used in the industries of the United States. Measured in terms of value, Germany manufactured 74 per cent of the world's output of dyes.[48]

In 1914, a large proportion of the dye factories in England and France were owned and operated by German dye concerns because the patent laws of these two nations required the "working" of patents. On the other hand, the Germans owned practically no dye factories in the United States owing to the absence of such legislation. During the war, France and England took over the German dye factories, but the United States had no such factories to utilize—a situation due to the suppression of United States patents by foreigners.

b. *During World War I.* A startling experience which reflected the German use of United States patents to hold back certain industries in this country was Germany's threat to withhold her drugs and dyes unless President Wilson was instrumental in lifting the English blockade on such things as cotton and oil. As F. D. Garvan of the Chemical Foundation testified in 1935, "Our very patent situation was allowed to be used by a foreign nation to become a club against our maintaining neutrality."[49] He stated that the experience in World War I, "showed that our chemical development had been seriously neglected by our people and that control of our chemical development had fallen into the hands principally of Germany; that they had not manufactured here, but they had used their patents here to crush out any advance of the American development."[50] Witness, for example, our inability to get started in the synthetic acetic acid field in 1917.[51] In 1924, the Federal Trade Commission reported that in 1917 "it found an acute condition as to certain synthetic drugs which had formerly been entirely of German manufacture and production."[52] Said President Wilson to Congress in his annual message of May 20, 1919, "Our complete dependence upon German supplies before the war made the interruption of trade a cause of exceptional economic disturbance."

c. *Between the First and Second World Wars.* To overcome its difficulties the United States sought to encourage the chemical, dye, and

48 *Dyestuffs for American Textile and Other Industries,* U. S. Department of Commerce *Special Agent Series No. 96* (1915), 30.

49 Pooling Hearings, Part 1, 746.

50 *Ibid.,* 740.

51 Bone Hearings, Part 3, 1365.

52 *Ibid.,* 1211.

drug industries here by making the confiscated patents of Germans available to domestic manufacturers and by passing a high tariff to foster the growth of infant industries in those fields. However, according to the alien property custodian in 1945, "In the interval between the two world wars a number of firms which had been seized by the Alien Property Custodian in World War I passed again into German hands. Altogether there are twenty-eight cases involving seizures of business enterprises in World War II which are related to seizures of business enterprises in World War I. In fourteen instances the seized company found its way back into the hands of the original owners. In ten instances, although the Germans from whom business enterprises were seized in World War I had not regained the firms taken from them, they had become the owners of interests in other American business enterprises, ordinarily in the same field of activity as their original company. One company, interests in which had been seized in World War I, was found again to be partly German-owned in World War II, but different Germans were the owners. In three other instances neither the companies nor the owners were the same in the two wars, but some connection existed between a firm taken in World War I and another taken in World War II."[53]

One factor that enabled Germany to re-establish its dominance in chemicals, dyes, and drugs was the importance of their postwar inventions. In one case, for example, it got back a block of stock in exchange for new patent rights obtained here. The mere seizure and sale of German patents could not make companies here entirely independent of German research and know-how.

The emergency tariff of 1922 was partly for the purpose of building up the chemical and other industries essential to war. These industries were to be a part of a preparedness program just as much as a bigger army and navy. However, the tariff, in order to prevent any maladjustment to industries already dependent on chemicals or other materials from abroad, provided that chemical intermediates might be imported at lower duties if not available in the United States. New intermediates were not made here because German concerns controlled the patents upon some of them—patents previously "confiscated" by the United States and later acquired by the Germans. They were manufactured only in Germany and were therefore obtain-

[53] EGRW Hearings, Part 4, 587.

able only from that country. Loopholes like this one enabled Germany to defeat largely the purpose of this tariff and therefore remain dominant in chemicals and dyes.

In the general revisions of the tariff in 1922 and 1930 the duties on products generally, including pharmaceuticals and dyes, were increased, but the Germans still retained their patent control of the most recent developments in these fields.

In 1935, Garvan said that the Germans were rapidly reinstating their position of patent control. "They have a way of patenting everything, whether they are ready to manufacture it or whether they see it in the advance realm of science."[54] He referred to the German patent on the idea of insulin; although undeveloped it was cited in opposing the insulin which became available to diabetics through the research and development of others.

According to a confidential letter of 1940 from Farben to the Reich Economic Ministry which came into the possession of the United States after the close of the war, 82 per cent of the organic aniline dyes were produced in Germany until the outbreak of World War II. "Germany's actual position in the world's dyestuff production is, however, not fully expressed by the share of 82 per cent, because the German dyestuff factories were forced by France, England, and Russia, through corresponding patent and customs legislation, to conduct part of their production in those countries. This was accomplished through the establishment of branch factories the production of which amounted, for 1931, to approximately 6 per cent of world production. Germany's share in the world's dyestuff business amounted thus to almost 90 per cent."[55]

In some instances the Germans established or reacquired subsidiaries in this country, notably General Aniline and Film, American Bosch Corporation, and the Schering corporations, to avoid the high tariff. However, either by restricting the important know-how to a few persons, or by incomplete disclosure in patents, they still retained dominance in these fields. Moreover, Farben in particular manipulated cartel agreements with Standard Oil of New Jersey and other American companies to the political as well as the economic advantage of Germany. Especially significant was Farben's control over strategic

[54] Pooling Hearings, Part 1, 745.
[55] EGRW Hearings, Part 10, 1417.

supplies for war. Some of this control was accomplished largely by Farben's suppression of United States patents. "Thus, in an international patent licensing arrangement between Rohm and Haas, a United States concern, and I. G. Farben, Rohm and Haas received patent licenses under which it could use acrylic acid and methacrylic acid for certain purposes, but not to make photographic articles, dyestuffs, artificial rubber, pharmaceutical articles and abrasives."[56]

d. *During World War II.* The failure of chemical industries to become established here as had been expected, the continued research and patent program of Germany, and the machinations of cartels found the United States under a big handicap in World War II as well as in World War I. The United States was deprived of crucial products in time of crisis, not because of lack of resources, but because of exploitation of our patent system. We unwittingly permitted the suppression of inventions and the creation of shortages. For ten years Farben's buna patents for making synthetic rubber were on file in Washington—without a pound of this product being made under them in this country. Even after Farben transferred these patents to Standard Oil of New Jersey in 1940, it withheld the know-how as the Nazi government had requested.[57] In the meantime, however, Farben was well informed about synthetic rubber developments in this country. Also there were shortages in drugs and dyes, and cartel agreements held down the output of magnesium, beryllium, and optical and other goods. A crucial drug which Germany controlled by patent was atabrine, a substitute for quinine; eventually, however, intensive research in this country, in conjunction with the confiscation of the German patent, made it possible to manufacture this substitute in sufficient quantities to overcome the scarcity of quinine from bark.

With the advent of the war, the United States seized enemy patents once again. By December 31, 1944, the alien property custodian had acquired 25,566 patents and 3,906 patent applications, unpatented inventions, and part interests in patents, all formerly owned by Germans.[58]

e. *After World War II.* Now that West Germany has acquired its freedom to make its own policies, will it employ patents, especially

[56] *Restrictive Business Practices,* United Nations Economic and Social Council *Report,* 13.

[57] Bone Hearings, Part 6, 2633–34.

[58] EGRW Hearings, Part 4, 593.

in the United States, to rebuild its economic and political strength? It will have a greater need than ever for dollars and therefore for research and patents and markets. So far as the United States goes, Germany's opportunity is still wide open. We have made no changes in our patent laws to prevent the suppression of our patents and the exploitation of them in other ways—all effects which we have known, tolerated, and endured for several decades.

The patents seized during the last war by the alien property custodian were transferred in 1946 to the Office of Alien Property of the Department of Justice. Those patents which are unexpired are still held pursuant to legislation precluding the return of them to their former owners. These patents are licensed royalty-free to nationals of the United States and of twenty-six other countries in accordance with international agreement, provided prewar licenses in good faith to a United States company did not preclude further licensing. Moreover, the Office of Alien Property has required vested companies to license their patents on nonexclusive reasonable-royalty terms. However, Schering resisted the compulsory licensing of its patents in accordance with an agreement of January 1, 1952, whereupon the Department of Justice brought suit against it in December, 1952. Also the Office of Alien Property sold the stock of the Schering and American Bosch companies upon the condition that if the stock ever got back in German hands it would revert to the government. Most of the stock of General Aniline and Film cannot be sold by the government because of a suit to recover it filed by the Interhandel Corporation of Switzerland.[59] Is this corporation a Farben dummy to conceal German ownership or control and prevent seizure of properties? The I. G. Farben of old, now broken up into smaller companies, no longer exists. Will another one be created in the future? West Germany's participation in the general scheme of free world defense may change its political objectives, temporarily at least. Will its urge for conquest of markets and governments assert itself again?

Legal Interpretation

As already noted, the legal rights of the patentee to maintain the resale prices of his patented product and to control the supplementary

[59] Department of Justice, Office of Alien Property *Annual Report* (1951) 43–49; and correspondence.

supplies for it were first upheld in the lower courts and later disapproved by the Supreme Court. The judicial history of the suppression of patents is somewhat the reverse. At least two lower courts condemned the practice before other courts, including the highest tribunal, gave their approval with reservations.

Hoe v. *Knap*. This decision, in 1886, involved the nonuse of patented machines due to the greater profitableness of the old machines and the expense of displacing the latter by the former. A circuit court stated that a patentee is "bound either to use the patent himself or allow others to use it on reasonable or equitable terms."[60]

Ewart Manufacturing Company v. *Baldwin Cycle-chain Company*. Another circuit court, in 1898, declared "that a patent for an invention which the patentee refuses to make available himself, and refuses to allow others to make useful, is not within the spirit of the provision of the Constitution which assigns as a reason for securing exclusive rights to authors and inventors a desire 'to promote the progress of science and the useful arts,' and that patents so held are entitled to scant recognition at law, though necessarily to some, but to none whatever in equity."[61]

Button-fastener Case. A circuit court of appeals, in 1896, rendered the first judicial decision which definitely upheld the legal right to suppress patents. It stated that the public has retained no other security than the self-interest of the patentee to insure the practical use of the invention. "A suppression can endure but for the life of a patent, and the disclosure he has made will enable all to enjoy the fruit of his genius. His title is exclusive, and so clearly within the constitutional provisions in respect of private property that he is neither bound to use his discovery himself, nor permit others to use it."[62]

Paper Bag Case. A decision very similar to the preceding one, except for the dissenting opinion of the lower court, was the paper bag case. Continental Paper Bag, the owner of a patent on a machine for manufacturing paper bags, acquired a competing patent and then suppressed it. A suit arose from the infringement of the unused patent. One of the arguments of the defendant was that the patent had been suppressed.

[60] 27 Fed. 204, 212 (C.C.N.D. Ill. 1886).

[61] *Ewart Mfg. Co.* v. *Baldwin Cycle-chain Co.*, 91 Fed. 262, 265 (C.C.D. Mass. 1898).

[62] *Heaton-Peninsular Button Fastener Co.* v. *Eureka Specialty Co.*, 77 Fed. 288, 294–95 (6th Cir. 1896).

In a circuit court of appeals two of the three judges decided that, in spite of suppression and nonuse, the complainant still enjoyed the right to restrain infringement by injunction.[63] The third judge registered a strong dissenting opinion. He said, "Not withstanding infringement, I contend that injunction relief should not be granted because it is an infringement of a paper patent deliberately held in nonuse for a wrongful purpose."[64] He conceded that, "standing alone, nonuse is no efficient reason for withholding injunction," and therefore stressed the intent and result of nonuse rather than nonuse per se.[65]

Upon appeal to the Supreme Court, Continental made at least a plausible justification of the nonuse. This tribunal, in 1908, stated: "It is certainly disputable that the nonuse was unreasonable or that the rights of the public were involved. There was no question of a diminished supply or of increase of prices, and can it be said, as a matter of law, that a nonuse was unreasonable which had for its motive the saving of the expense that would have been involved by changing the equipment of a factory from one set of machines to another? And even if the old machines could have been altered, the expense would have been considerable. As to the suggestion that competitors were excluded from the use of the new patent, we answer that such exclusion may be said to have been of the very essence of the right conferred by the patent, as it is the privilege of any owner of property to use or not to use it, without question of motive."[66] The court, it should be noticed, suggested a loophole to this sweeping statement by adding, "Whether, however, a case cannot arise where, regarding the situation of the parties in view of the public interest, a court of equity might be justified in withholding relief by injunction, we do not decide."[67] Justice John M. Harlan dissented from this opinion; he concluded that "the facts are such that the court should have declined, upon grounds of public policy, to give any relief to the plaintiff by injunction."[68]

Lock Case. A patentee may not agree to suppress his patent for the purpose of restraining trade in his patented commodity. In the lock

63 *Continental Paper Bag Co.* v. *Eastern Paper Bag Co.,* 150 Fed. 741 (1st Cir. 1906).

64 *Ibid.,* 744.

65 *Ibid.,* 745.

66 *Ibid.,* 210 U.S. 405, 429 (1908).

67 *Ibid.,* 430.

68 *Ibid.,* 430; this decision is referred to with approval in other cases—261 U.S. 24, 34; 263 U.S. 50, 55; and 323 U.S. 386, 432.

case a circuit court, in 1909, declared, "Granting that nonuse of an invention is fully within the right of the owner of a patent, it does not follow that he may by agreement bind himself to nonuse, save in connection with an assignment of his letters patent. Ownership of a patent involves no obligations to use, nor does ownership of other property. Nonuse ordinarily violates no law; but contracting with another, putting it in the power of another to compel one not to use, is a contract in restraint of trade, designed for the purpose of suppressing competition."[69] This case apparently is an early instance of refusing to enforce a patent on grounds of misuse.

Vitamin Case. In 1944, a circuit court of appeals, in passing upon the right of the owner of patents on irradiation to withhold the invention from manufacturers of oleomargarine, pointed out that the right of suppression "has not been raised in connection with the public interest in restoring the health of the afflicted." It then quoted the loophole in the paper bag decision already noted.[70]

Special Equipment Company v. *Coe*. Another circuit court of appeals, in 1944, decided that the commissioner of patents could not be compelled "to issue a patent upon an application for a sub-combination of the elements of a machine for which the inventor had previously filed a patent application" upon the ground that the applicant "did not intend to make or use the invention and that the purpose of seeking the patent was to exploit and protect the combination invention embodied in the complete machine, of which the sub-combination is a part."[71]

In reversing this stand by a five to four vote the Supreme Court, after declaring that it "has consistently held that failure of the patentee to make use of a patented invention does not affect the validity of the patent,"[72] stated, "We have no occasion to consider here . . . whether the courts on equitable principles should decline to enjoin patent infringements or decline to compel the issue of a patent if and when it appears that the patentee or inventor intends to make no use of the invention."[73]

[69] *Blount Mfg. Co.* v. *Yale and Towne Mfg.* Co., 166 Fed. 555, 560 (C.C. Mass. 1909).

[70] *Vitamin Technologists* v. *Wisconsin Alumni Research Foundation,* 146 F. 2d 941, 946 (9th Cir. 1945).

[71] *Special Equipment Co.* v. *Coe,* 324 U.S. 370, 371 (1945), reversing 144 F. 2d 497 (D. C. Cir. 1944).

[72] *Ibid.,* 378–79.

[73] *Ibid.,* 379.

Three dissenting justices—Douglas, Black, and Murphy—declared, "The right of suppression of a patent came into law over a century after the first patent act was passed"[74]—in the button-fastener and paper bag decisions—and then added, "It is time to be rid of that rule. It is inconsistent with the Constitution and the patent legislation which Congress has enacted."[75] More specifically, the dissenting justices held that the court "should withhold its aid from a patentee who has employed or plans to employ the patent not to exploit the invention but to suppress it in order to protect another patent or otherwise."[76] Justice Rutledge also dissented but on a different ground.

A bare majority, five to four, and the intimation that a decision on "equitable principles" might be different indicates that the Supreme Court may be on the verge of restoring the obligation of a patentee to promote the progress of science and the useful arts by using his invention.

Criticism

The criticism of suppression is presented in terms of the historical background of our patent system, of use versus description in the advancement of technical arts, of discard or retention of old equipment, of competition in alternate inventions, and of conflict between national and foreign interests.

Historical Background of Our Patent System. The whole historical background of granting monopolies for promoting the progress of the useful arts gives no sanction to the suppression of inventions. Even prior to colonization in America, England had promoted industry by allowing inventions to be brought in from abroad, sometimes by foreigners, and by recognizing inventions resulting from new processes and products. But whatever their origin, inventions had to be put to use, not just described, if they were to be patented. The period was for fourteen years—long enough for two sets of apprentices to learn the art and then practice it throughout the realm. The result of England's patent system was new industries, such as weaving and glass making, and this experience of using patents to promote indus-

[74] *Ibid.*, 380.
[75] *Ibid.*, 381.
[76] *Ibid.*, 384.

try was part of the background of the people who settled along the eastern shores of North America. After these English colonial settlements became organized and felt the need for salt, glass, and other essentials, they too offered temporary monopolies to those who would provide such products. In some instances the grant specified the minimum amount of production that was to take place and the time within which production was to begin. Connecticut, in 1747, granted the "sole liberty and privilege of making glass in this Colony" for twenty years provided that at least five hundred feet be made during each of the last sixteen years.[77] Note an allowance of four years to get ready for production. As a rule the production could be by the patentee or his apprentices. Massachusetts, in 1750, granted a patent to a manufacturer of sperm candles, "requiring him to teach at least five apprentices during that period, of whom two should be nominated by the General Court."[78] By 1750, patents were restricted to new inventions and therefore did not include old arts brought in from abroad.

The granting of patents by the colonies for inventions—still upon the conditions that they be introduced—continued until the adoption of the Constitution. In 1780, just one year before the adoption of the Articles of Confederation, Pennsylvania granted Henry Guest "the sole and exclusive right, for the term of five years, of manufacturing oil and blubber from the materials he has discovered," to become effective provided the grantee within eight months build a factory and produce the product within the state.[79] In 1784, South Carolina enacted apparently the first real general patent law, "An Act for the encouragement of arts and sciences," and it provided for compulsory licensing in case of suppression.[80]

The forefathers who adopted the Constitution knew about this practice of granting temporary monopolies to those who make their inventions available in products. With the experience of the colonies in mind they authorized Congress "to promote the progress of science and the useful arts by securing for limited times to authors and inventors, the exclusive rights to their respective writings and discoveries." It was adopted at a time of antagonism to monopoly and at a time of emphasis on individual rights—even fear of unauthorized en-

[77] Pooling Hearings, Part 1, 596.
[78] *Ibid.*, 597.
[79] *Ibid.*, Part 4, 3570.
[80] *Ibid.*, 3570–71.

croachments which might occur upon liberty brought about the first ten amendments to the Constitution. But there was certainly as great a need for the introduction of new products in the nation then as there had been in each colony, and granting temporary monopolies to those who would make, use, and sell inventions rather than just describe them was in keeping with what England and the colonies had done and with the needs of the times.

In defending the Constitution, in 1788, James Iredell, later one of the first associate justices of the Supreme Court, said, "The interest of the proprietor (of a copyright) will always induce him to publish a quantity fully equal to the demand."[81] The same statement, of course, would apply to a patent. This statement of his was designed apparently to quiet the fears of possible suppression. A commissioner of patents, James T. Newton, testified, in 1919, "And the value of the patent is right there; it enables the manufacturer to get an industry started. That is the value to the public, too. After the industry is started and the public gets the benefit of the invention, then the patent is beginning to perform the function that the framers of the Constitution intended it should perform, in making work easier for the general public, in giving them something they never had before, and something that they wanted."[82]

It is logical that conferred rights assume positive rather than negative action. "To promote the progress of science and the useful arts by securing for limited times to authors and inventors, the exclusive rights to their respective writings and discoveries," takes for granted the publication of the writings and the practice of the discoveries rather than the suppression of them.

Use Versus Description in Advancement of Technical Arts. The spirit of our patent system is diametrically opposed to the suppression or "nonworking" of patents in spite of the fact that inventions so treated may be disclosed in patents issued by the Patent Office. The public, in keeping with the philosophy and purpose of the patent laws, should expect more than a mere description of the invention in return for the exclusive monopoly granted to the patentee for a period of seventeen years. As one writer has said, "Now, although full and honest disclosure of an invention constitutes the lawful consideration

81 *Ibid.*, 3572.

82 Pooling Hearings, Part 1, 601.

which will support the inventor's right to a patent, it is well known that such disclosure on the part of the inventor is not, in many cases, a consummation of the great object of the patent laws, namely, actually to place the patented improvements upon the market and thereby actually demonstrate the merits of the improvements and develop any possible latent defects."[83] R. H. F. Palgrave once said, "The granting of a patent may be a hindrance to industrial life if the patentee, through want of energy or want of means, neglects to work his invention, whilst others, who would be willing and able to do so, are afraid of infringing his rights."[84] The most efficient and profitable way in which the people can learn of an invention is by using it or the product which it manufactures. Moreover, the inventor is encouraged psychologically by seeing his invention used.

The expectation that the patentee will use his invention is implied in our infringement laws, which protect the legal monopoly of the patentee—the right to exclude others from making, using, and selling his invention—throughout the entire period during which his monopoly extends. The corollary of this protection, it would seem, is that the patentee must develop his invention.[85]

The need for putting an invention into practice so that others may use the process or make the product is more necessary now than in the past because of the increasing extent of incomplete disclosure. Many patent owners have not adequately described their inventions in their patents. The German patents on synthetic rubber, taken out and suppressed in this country, hindered those who sought to produce synthetic rubber by kindred means but could not until the seizure of the German patents by this government and until American research revealed the techniques which the German patents failed to disclose.

The alleged right to suppress an invention is based on the assumption that an inventor in obtaining a patent fulfills his part of the contract by disclosing his invention. According to this argument, a patent is analogous to a farm or house or other physical property in that the owner has the right to use or not to use and in the particular way he pleases. This fiction ignores the history and purpose of patents

[83] 67 Cong., 1 sess., *Senate Doc. No. 6,* 43.
[84] *Dictionary of Political Economy,* III, 76.
[85] William C. Robinson, *The Law of Patents for Useful Inventions,* 65–67.

and the impossibility of disclosing chemical and electrical inventions, especially if the applicants seek to withhold and hide rather than reveal. The best description is no substitute for use either in learning about the invention or in satisfying human wants. The legal theory of ordinary contract, applied here, fails to conform to the history and purpose of patents.

Discard or Retention of Old Equipment. Another question that arises is whether it is socially desirable to suppress patents in order to utilize old equipment more completely. It is argued that the immediate adoption of a new invention, such as the automatic telephone, however worthy, may mean maladjustment and unemployment of labor and waste of capital and, therefore, that it is desirable to delay or graduate the introduction of the invention. This contention undoubtedly has considerable merit. But it should be remembered that technical and industrial progress may be measured by the size of the scrap pile. Witness, for example, in the technical development of the American railways how engines and cars were discarded and more efficient engines and cars adopted. The abandonment of the old necessitates considerable waste; the adoption of the new usually means progress. Each is a balancing factor to the other, and the one that predominates depends largely upon the circumstances of each particular case. In any instance it should be borne in mind that the suppression of patents per se, whatever the excuse, discourages invention and falls short of fulfilling the spirit of the patent law and the expectations of the public, and that competition, one of the cornerstones of the existing economic order, implies the adoption, even with attendant wastes, of the most efficient methods by gain-seeking rivals.

Competition in Alternate Inventions. The single ownership of alternate inventions and the suppression of all except one of them prevents rivalry in processes and methods as well as in prices and, therefore, violates the ideal of competition which underlies our economic system. The United States abounds with illustrations, both historical and current, of the policy and practice of competitive monopoly or monopolistic competition. The transportation field affords excellent illustrations. A charter or franchise was granted to build a canal, a turnpike, or a railway over a particular route—not over all possible routes, with a choice of one of them, presumably the best, and the nonuse of the others, and not with the opportunity of using one route and buying up the others. Likewise, in more modern times, a permit to an airline does not cover all possible routes, with the privilege of using one of them and keeping out competition from the others. Incidentally, building and operating different kinds of transportation facilities ahead of economic development of the country and ahead of the demand for their services was fraught with financial and other hazards. Each canal, road, railway, or airline had its own territory, with some degree of monopoly and also the competition of rivals. Each railway, subject to government regulation, has had a degree of monopoly in local traffic but has had competition with rivals between cities served by two or more roads; all railways must compete with trucks, ships, pipe lines, and planes. Likewise alternate inventions within each of these fields have competed with each other.

Who denies the beneficial effect of all these phases of competition in developing transportation facilities in the United States? As previously explained, much less competition in communication accounts for preventing and retarding the introduction of improvements in this field. Also in manufacturing, competition has been checked by one company getting control of the patents in its field and then suppressing certain inventions until others have been fully exploited. The policy of United Shoe Machinery, for example, meant the freezing of technology in shoe machinery already built but not worn out. Additional illustrations are found in other patent consolidations and in patent pools, already described.

On the other hand, patents and competition are found in the oil-field equipment and many other industries, and much of it has come from the comparatively small and new company. The president of a core-drilling company stated that competitors "found other means of getting around the patent we had, but in spite of that fact we have been able to progress during this time."[86] The manufacturers of a patented oil screen testified, "We have four competitors in our trade territory who make a similar article but under separate patents."[87] In other words, each product was both monopolistic and competitive, protected by patents and yet only one of five for the same purpose. Note that in these and subsequent examples, rival inventions are made and sold and used in competition with each other, and there is no oc-

[86] Comp. L. Hearings, 164.
[87] *Ibid.*, 82.

casion for suppression. According to a representative of the Rubber Manufacturers Association, "Our patent system stimulates progress, not only in providing for the inventor the possibility of a reward based on reasonable protection of his invention," but also "in forcing competitors to seek different ways to accomplish a patented result."[88] The chief engineer of a company with 140 patents on machinery for the hydraulic transmission of power, said, "Nearly all patented machines and processes have competition either with another patented machine or with an established machine or process which are so old that the patents have expired. All that the ordinary patent does is to give the inventor and investor in a new device some assurance that when their work is successfully completed and on the market it cannot be directly copied by an aspiring competitor," who "must invest his own money and do his own thinking as a basis of competition."[89] George Baekeland, of the Bakelite Corporation, testified, "Now, we have a number of competitors. They all do research work. Sometimes they turn out a product that is a little better than ours, or we turn out one better than theirs, and as soon as that happens everybody goes to work and tries to get a material which will do the same thing, or better it—probably to better it—and there is a great deal of rivalry and competition for that reason."[90]

An inventor and manufacturer of machine tools said, "We find our competitor moving ahead a notch with some new patented improvement, and we have to think . . . of something else that is better than that. . . . The competitor is just about as good as we are, and we are a continual stimulus to each other. That doesn't trouble us at all, and it is good for the industry as a whole."[91] Contrast this situation with that in the shoe machinery or telephone business with a policy of seeking all patents in the field and suppressing many of them to prevent competition and avoid obsolescence. A manufacturer of automobile wheels stated, "A patent granted to a competitor in this industry has proved to be the greatest incentive possible to other competitors. For example, if one of my competitors tomorrow should bring out a wheel which would revolutionize the wheel industry and threaten to put the company with which I am connected out of busi-

88 *Ibid.*, 169.
89 *Ibid.*, 270.
90 *Ibid.*, 348.
91 TNEC Hearings, Part 3, 929.

ness, that in itself would be the greatest incentive in the world for us to use every possible means to get around that patent and to devise something quickly to save our very lives and our physical existence, and I am sure we would do it. We have been faced with that situation time after time."[92] The same rivalry has existed in developing automatic transmission devices. There are several such devices, and there has been a constant effort to improve them, thanks to competition.

If a company or person is charged with infringement, the first step usually is to decide whether such encroachment exists and whether the infringed patent is valid. If affirmative answers result, the next step is to go to the research staff and ask them to make a device or invention that avoids the patent in question. The outcome is frequently a better and cheaper product. Such a situation stimulates small companies and therefore wholesome competition.

These illustrations, and there are others, show that patents and competition exist in many industries. This country from its very beginning has been committed both to technical progress and to competition—both to patented inventions and to rivalry. According to this commitment every company may have its own legal monopoly, each shall be in competition with the others—a dual situation in harmony both with the patent law and with antitrust legislation. On the other hand, the concentration of all these alternate patents in one company and the resultant suppression of patents and restraint of trade violate the spirit if not the letter of both laws.

Conflict of National and Foreign Interests. The suppression of important United States patents by individuals and by business concerns of this country results in an even worse situation if the individual and business concern do not secure corresponding patents in other nations. It means that the people of the United States are forbidden to make the inventions patented here, while foreigners are free to develop them abroad.

The arguments against suppression are accepted most readily when attention is drawn to those patents, granted by the United States to foreign citizens, which are used to benefit other nations. Our laws allow foreigners to take out patents in this country merely for the purpose of reserving the United States as a market for their patented products; the result in some instances is the prevention of manufac-

[92] *Ibid.*, 1050.

ture in this country in spite of more favorable factors of production here than abroad. The only benefit arising here comes mainly from the importation of such inventions or the products which they make possible, and even this may cease in time of war or other crisis.

CHAPTER IX

Discouragement of Inventors

IN APPRAISING the effect that the patent system has on invention, it is necessary to differentiate two general classes of inventors—the independent and the hired. Hired inventors in turn may be divided into two main groups—those employed by business corporations and those employed by universities, the government, and other nonprofit organizations. In an earlier day, the independent inventor predominated but his importance has diminished as the economy has shifted from the artisan and his small shop to the corporation with its factory and laboratory and to nonprofit organizations with their research projects. Today's inventor is most generally remunerated not by the sale of an invention or patent rights but by the receipt of wages or salary for his labor and for his inventive ability.

Independent Inventor

If the inventor sells his patent rights at all, it is usually for a lump sum rather than for royalties. If he develops his own invention, which is seldom, he must seek the capital of others. Most of his inventions are never sold or developed at all. In any case he usually receives little or nothing. As the obstacles of the inventor have grown, patents, to an increasing extent, have stimulated him through delusion rather than reward. What are his difficulties? Why does he receive little or nothing for his efforts? For one reason, defects in the invention prevent success in many instances. Another reason is that the inherent characteristics of a patent system may prevent an inventor, however worthy his invention, from obtaining any reward. Other obstacles involve the defects of our patent system—inefficient and unethical attorneys and promoters, interference proceedings, litigation, lack of capital, lack of market, fencing in, and unfair competition.

Defects in the Invention and of the Inventor. Independent inventors generally lack scientific training, pertinent information, and practical guidance. Consequently they patent many ideas on worse ways of doing something that is well done already. Even the best inventions must be modified in most instances before they can be utilized.

The disappointment of the inventor, and especially of the great inventor, may arise from his personal characteristics. The mechanical genius is not likely to be a capable organizer and manager; the great inventor, like the artist, is usually so absorbed in his thoughts that he cares but little for material gain. In fact, an inventor may be eccentric because of the very qualities that make him an inventor. He is apt to overestimate the importance of his invention and therefore to become the victim of bitter disappointment when his exalted hopes are shattered. The temperament of the inventive genius yearns for a just valuation of his contribution to the social welfare, but frustration rather than appreciation results when business leaders like Cornelius Vanderbilt dismiss inventors like George Westinghouse and inventions like the air brake by saying there is no time to waste on fools.

Characteristics of Inventions and Patent System. Invention represents a growth, a prolonged and persistent effort by many individuals. There is a period of conception, a period of development, and a period of achievement, and they may be separated by many years. Those who conceive an invention seldom receive their fair share of recognition and remuneration. In fact, an invention may be so far ahead of the times that the term of the patent may completely expire before the public fully understands, adopts, or appreciates the invention.

No matter how independent and original and worthy an invention may be, its author may fail to secure a patent or any other award because somebody else, who has really created no more, is ahead of him. He receives no second or third prize, perhaps not even honorable mention. In fact he is even excluded for a period of years in practicing his own invention.

Existing patents, however worthy in stimulating invention, are at the same time "verboten" signs that confront the inventor. It is a common practice, especially of large companies well financed and equipped with technicians and patent lawyers, to take out every possible patent in their fields and thus block any would-be intruder. If an outsider seeks a patent in this domain, he must find out in some instances about hundreds of patents on kindred ideas and avoid them.

Creative minds may be compelled to spend more time in obtaining or avoiding patents than in solving a problem.

Inefficient and Unethical Attorneys and Promoters. Some patent solicitors, though apparently limited to a small minority, strive for "quantity production"; in other words, they are more interested in securing fees than in giving proper service to their clients. They recommend the securement of the most trivial patents, those covering the most minute features or improvements. Moreover, a few patent attorneys, in an effort to rush patent applications, may agree to cancel claims to which objections have been made by the Patent Office and may, therefore, secure a patent which contains only narrow and insignificant claims, one that fails to protect the essence of the invention. It should be stated in this connection that the Patent Office objects to one or more claims of about 90 per cent of the applications for patents. Furthermore, this same small group of unethical attorneys, in an attempt to secure a large number of patents for clients, may recommend, in spite of the insignificance of the invention and foreign laws requiring annual fees of considerable amounts, that they take out patents in other countries.

Today, because of stricter rules of the Patent Office, bargain fees have been substituted for free searches. In 1952, the author, in response to advertisements of eight different patent attorneys in a well-known magazine, requested copies of the attorneys' proposals for securing patent protection. All of the attorneys sent printed forms for describing and drawing the invention. All eight also proposed a search of United States patents to determine the invention's patentability and their charges for this service ranged from five to fifteen dollars. Six of them offered another service, designated by such names as "special" and "advance," which included the search already mentioned and also prints of the drawings of the invention in case of a favorable report on patentability. The alleged advantage of the second plan was avoidance of unnecessary delay, and the fees varied from twenty-five to thirty-five dollars. One patent attorney, for example, asked the prospective client to choose one of these services:

1. I am sending herewith my Record of Invention form (or other disclosure) with $5.00. Please proceed with *SEARCH* of U. S. Patent Records, and report before proceeding further.

2. I enclose $25.00 with my Record of Invention form to make my case *SPECIAL* so you may proceed and send me Approval Prints of the Official Drawings of my invention, provided the Search report is favorable—thus saving TIME. A refund is due me if Report is unfavorable.

Service No. 1 was described in a form letter as follows:

Upon receipt of such a disclosure along with the nominal search and report fee of $5.00, I shall proceed with a search of the U. S. Patent Records and report to you your chances of securing patent protection.

The search of the prior U. S. Patents at the Search Room of the U. S. Patent Office is an important step outside of actual filing in the U. S. Patent Office. In my search report I will indicate your chances of obtaining patent protection in view of any patent cited by my searcher. If references are cited by my searcher, copies of the patents will be sent you.

When you have the search report in your hands, you can then determine whether you want to proceed any further.

It should be added that this circular on "Further Facts about Patents" included the statement, "Just as there is no substitute for a United States Patent, there is no other method so efficient and direct in securing patent protection as retaining the services of a competent registered Patent Attorney."

A follow-up letter from one firm—about two months after the first one—started off with "You can save $5.00 if you act soon!" and then stated that the fee for a preliminary search was reduced from fifteen to ten dollars. However, "This offer is good only if you take advantage of it within thirty days from date of this letter."

The advertising of the eight attorneys, whose bargain fees for searches have been described, emphasized *can* rather than *should* a patent be obtained. Such headlines as "Delays Are Dangerous in Patent Matters" and "Why Prompt Action Is Important" may result in getting a patent in haste and repenting at leisure. One lawyer enclosed a yellow sheet with the following heading and message:

RELATIVE TO CONTACTING MANUFACTURERS

I have formulated a plan for my clients whereby to interest manufacturers in acquiring inventor rights, after filing of application for patent.

Perhaps you will be interested in this plan, if and when, we file a formal patent application on your invention.

This appeal is a close approach to the former stories of fortune that awaits inventors.

Interference Proceedings. If two or more applications for the same invention are pending in the Patent Office at the same time, an interference is declared for the purpose of ascertaining the first or prior inventor. The interference may arise even after a patent has been granted because of a clash between its claims and those of a pending application. A board of patent interferences gives the first decision and then appeals may be taken to the courts. Another and subsequent interference may necessitate a repetition of similar proceedings. In such proceedings the poor inventor is at a great disadvantage. Unable to withstand the financial strain, he is often under pressure to compromise, and even surrender, to a "well-heeled" opponent.

Litigation. Litigation, which arises mainly from the invalidity of many patents, is a crushing burden, and it reduces the effective life of a patent.

Heavy expense results from the necessity of having searches made, hiring skilled attorneys, arranging expert testimony and documentary evidence, and making extensive preparation for trials. As a result the lone inventor has only a remote chance, if any, of obtaining any net financial return from a patent which he tries to maintain.

The president of Thomas A. Edison, Inc., once stated that Edison had spent more money in obtaining patents, litigating them, and preventing infringements of them than he had received from his patents as such.[1] More recent inventors, such as Lee De Forest, have won court decisions in behalf of their patents, only to find the expense pushed them into bankruptcy.

Litigation is at its worst when the government instructs contractors to proceed with production regardless of the patents of others and takes the responsibility for their infringements. The government, however, does not appraise the claims of patentees and make awards, as in England, according to the judgment of some board or commission. In practice the patentee is merely permitted to sue the government for damages in the Court of Claims; in September, 1953 there were fifty-two suits involving patents pending in that court.[2]

Lack of Capital. Although the cost of obtaining and defending a

[1] Oldfield Hearings of 1912, Part 2, 32.

[2] United States Code 28, Sec. 1498 as amended, and Code 35, Sec. 183. *Richmond Screw Anchor Co.* v. *United States,* 275 U.S. 331 (1928); *Bereslavsky* v. *Esso Standard Oil Co.,* 175 F. 2d 148 (4th Cir. 1949); *Bereslavsky* v. *Standard Oil Co.* (N.J., 82 F. Supp. 939 (D. Md. 1949).

patent may be nominal, it may still be true that a new invention requires an enormous amount of capital for development purposes. When the patent system was adopted more than a century ago, the inventor himself would often make and sell his invention. It was a time of relatively simple tools and devices, the production and sale of which required no great outlay of capital. However, modern methods of manufacturing and marketing a new product, and especially a complicated one, make the early practice of production largely inapplicable. The invention and the machinery for making it must be perfected and its introduction to the public requires salesmen and advertising.

Even if the patentee welcomes the co-operation of the capitalist and agrees to give him a fair share of any return from his invention, he may still be unable to secure sufficient capital for its development. Banks as a rule refuse to finance a new invention of any sort, for they cannot afford to lend money with mere possibilities as security. In some instances the lone patentee succeeds either in borrowing money from his friends and other individuals or in selling an interest in his invention to them. But the uncertainty of any new invention and the defects of our patent system deter most people from making a financial investment. The same reasons often explain the difficulty in securing remuneration from the outright sale of patents. Insufficient capital, to repeat, explains the failure of many patentees to secure any appreciable reward from their inventions.

The importance of capital, as indicated, tends to direct the efforts of the poor inventor along those lines which require a minimum of capital. The inventor of the adjustable shoe tree did much better financially than independent inventors of shoe machinery. Again, the man who invented and commercialized the flash light left a fortune for charities, while Goodyear, like several of the great inventors, went down to a pauper's grave.

Lack of Market. Pools, consolidations, license agreements, and cartels—any concentration in the control of patents—have deprived the inventor of a proper market for his invention. Even when such concentration was much less than it is today, President Wilson said, "I am not saying that all invention has been stopped by the growth of trusts, but I think it is perfectly clear that invention in many fields has been discouraged, that inventors have been prevented from reaping the full fruits of their ingenuity and industry, that mankind has been deprived of many comforts and conveniences, as well as the opportunity of buying at lower prices."[3]

Most pools are on a permanent basis and the inventions of each and all are shared alike. Such an arrangement, as noted in Chapter Two, makes it unprofitable for any particular member to acquire and develop the invention of any outsider. General William Mitchell, in 1935, expressed the opinion that the Aircraft Manufacturers Association is detrimental to the lone inventor. "That is already the case with a great many very able inventors. Some of them have gone abroad."[4]

Also, Richard E. Marine, who was in the Patent Office about twenty-four years and resigned from the Board of Appeals to join a private patent law firm, testified in 1938, "I have several clients myself at the present time with apparently very attractive devices, but they involve a change of standards and development work which the companies are not willing to undertake, largely because, if one company gets the invention and by research and development work commercializes it, the benefit of that will be given to the competitors, under their general plan of exchanging licenses for ordinary inventions."[5]

The record shows that monopolies, particularly patent consolidations, have refused in several instances to purchase important developments in their fields. The telegraph and cable companies refused to purchase the telephone for $300,000; all three of them declined to buy the wireless telegraph; and all four of them turned down the wireless telephone at $250,000. In reference to such refusals Louis D. Brandeis, a former justice of the Supreme Court, once testified, "These great organizations are constitutionally unprogressive."[6] According to Reginald A. Fessenden, "The greater the contribution, the more certain is it to be denied recognition by the entrenched corporations and their servile laboratory staffs. And the lack of such recognition coupled with the lack of properly developed models to convince by demonstration those who cannot understand technical presentations, explains the shameful spectacle of every single one of the

[3] MP Hearings, Part 2A, 349.
[4] Pooling Hearings, Part 1, 860.
[5] Comp. L. Hearings, 372.
[6] Oldfield Hearings of 1912, Part 18, 12.

world's great inventions having been forced to lie idle until outside competition had forced their adoption despite the cunning and conspiracy of the great corporations in that field—and often only after the inventor was no longer here to receive his due reward."[7]

An incidental but important phase of the refusal to purchase or adopt an outside invention is the feeling of those in charge of production and research that it would be a reflection upon them. As one inventor, Henry J. Gaisman, said in 1935, "They do not like to see an outsider come in with that invention which may embarrass them in their position. They should have thought of it."[8]

As just indicated, an entrenched business seeks equilibrium rather than upheaval in technology. An invention is not bought by a company unless the company's position is threatened. Its primary interest is to preserve and protect what it already has. It is safety-seeking rather than risk-taking. However, once a competitor threatens with a real invention because of the company's refusal to purchase or sell certain patents, the company becomes anxious to obtain patents for fencing him in.

In some instances the dominant concerns in an industry may have a purchase agreement covering patents. Perhaps the first important example of collusion in acquiring patent rights was the board of patent control maintained by GE and Westinghouse from 1897 to 1911. Both companies agreed that neither of them would acquire any license under any patent except by giving at the same time an option by which the other one should get a license on the same terms. Also, an agreement between GE and American Telephone provided that neither would acquire patents or secret processes without giving the other an opportunity to acquire rights for its field.[9] Likewise contracts between RCA and other members of the radiotelephone pool forbade any member to acquire patents or inventions applicable to the field of any of the others without giving them an opportunity to participate in the negotiations.[10] Again, according to a government complaint, four manufacturers of high-tension cables and accessories that cross-licensed each other agreed to share equally the expense of

7 Pooling Hearings, Part 1, 683.
8 *Ibid.*, 691.
9 *Ibid.*, Part 2, 1233.
10 *Ibid.*, Part 4, 3396.

obtaining pertinent patents, inventions, and technology from others and therefore not to compete against each other in acquiring them. As indicated in Chapter Two, a district court, in 1948, enjoined these companies from any agreement to refrain from competition in obtaining rights to patents and technological information.

License agreements often require the licensee to assign its patents to the licensor. Such an arrangement lessens interest in outside inventions and yet builds up the principal or only market for the patents of an outsider.

A cartel even may be built upon a pool, consolidation, or license agreements—any one, two, or three of them formed on an international basis, would constitute a cartel—and thus may remove any competition on a world-wide basis in buying patents. In the cartel agreement between Du Pont and ICI, the two companies "undertook that they would not purchase patents for each other's territory except after having given the other the privilege of pre-emption on the same terms as those under which the patent was offered to them."[11]

The handicap of the inventor in selling his patents was serious enough even in 1912 that President Taft referred to it in a message to Congress by stating, "Large corporations, by absorbing patents relating to particular arts, have succeeded in dominating entire industries, and the only market to which an inventor of improvements upon such machines may offer his patents for sale is to such corporations."[12]

Fencing In. A patentee may be deprived of the value of his invention by other patents that obstruct his invention. Some corporations maintain patent experts and technicians to anticipate possible improvements in competing inventions, secure patents upon the improvements and thus make it difficult, if not impossible, to develop and market them. In certain instances the owners become either licensees or leave the field. Earlier chapters have indicated that various corporations have indulged in the "fencing in" scheme in order to get rid of competition.

According to a Hartford-Empire memorandum of 1930, "We now

[11] *United States* v. *Imperial Chemical Industries, Ltd.,* 100 F. Supp. 504, 557 (S.D. N.Y. 1951).

[12] Pooling Hearings, Part 1, 1076.

have a number of applications which were filed to definitely forestall the development of competing machines of others."[13] At a TNEC hearing the president of Hartford-Empire responded as follows:

> The Chairman. Let me put it this way: you do watch these competing machines, do you not?
>
> Mr. Smith. Yes, we do.
>
> The Chairman. And in your research laboratory you study them for the purpose of developing improvements upon them?
>
> Mr. Smith. Right.[14]

How may the independent inventor overcome the patents that fence him in? As just indicated, he may take out patents which block or keep others out of his way, but ordinarily he cannot do this as quickly and thoroughly as a corporation with a corps of inventors and technicians and lawyers. He may, after finding out that certain patents fence him in, challenge their validity—and a high percentage of adjudicated patents are held invalid—but seldom does the inventor do so because of the expense. Public domain thus wrongfully fenced in usually remains private preserves. The lone inventor may even try to invent around the patents which fence him in; the handicap that other patents present may sharpen the wits—but the price is high. Too frequently, frustrated and discouraged, the lone inventor abandons his invention altogether.

According to a circuit court of appeals, "The result of granting blocking or fencing patents is to create a maze of patent restrictions whose effect is to confuse and impede business competitors and inventors and to entrench some one corporation in the position of domination over an industrial technique."[15]

Unfair Competition. The use of patents as a weapon for unfair practices, though not so evident now as in the past, still exists in isolated cases. The most effective scheme has been litigation, threatened or actual, initiated by the strong to force the weak to go bankrupt, to sell their patents, or to become licensees.

a. *Threats of Infringement Suits.* Such threats are made in bad faith if the company making the threat subsequently refuses actually to start and finish legal proceedings to determine the question of in-

13 TNEC Hearings, Part 2, 777.

14 *Ibid.*, 393.

15 *Special Equipment Co.* v. *Coe,* 144 F. 2d 497, 499 (D.C. Cir. 1944).

fringement. The effect of such threats upon the patentee or his company can be readily understood. The vice-president of Owens-Illinois, after asking a small manufacturer of melting and annealing equipment to join a proposed combination, said to him, "If you don't go in with us on this thing we will enter suit against you and we will continue to sue you until you are out of business."[16] A district judge referred to "threats of litigation backed by the financial power" of Hartford-Empire and its allies.[17] In another case, the promoter of the monopoly of parking meters, after stating that relevant patents might be invalidated in the courts, declared, "At present they are accorded a *prima facie* validity which could be used to advantage in discouraging competition."[18] A consent decree of 1946 enjoined the Corning Glass Works from excluding others from the market by threats of patent infringement suits.[19]

General Electric and its tungsten-carbide licensee threatened to bring infringement suits against small competitors. According to a district judge in 1948, "The striking feature of the record is that, although the first suit was filed in 1930, the patents were not judicially considered until 1940."[20] The judge pointed out that "the delay in reaching an adjudication of the patents was due to a number of reasons: In some instances, GE deliberately refrained from instituting suit. On one occasion suit was dropped for the reason that the defendant discontinued business. On another occasion the litigation came to an end when the defendant took an agency license."[21]

The attorney for General Development Company, a pool for variable condensers, reported the following to his client: "I also believe these patents may be useful in helping to get American Steel Package Company into the fold. After Variable's agreement is signed, I will recommend my again approaching American Steel Package Company charging them with infringement of other patents and possibly with the Cramer patent, in an endeavor to get American Steel Package to

16 TNEC Hearings, Part 2, 599.

17 46 F. Supp. 541, 618 (N.D. Ohio, 1942).

18 *United States* v. *Vehicular Parking,* 54 F. Supp. 828, 839 (D. Del. 1944).

19 *United States* v. *General Electric Co., Corning Glass Works, et al.,* CCH 1946–47 Trade Cases ¶ 57,448 (D.N.J. Civil No. 2590; Complaint, 1942; Consent Decree *re* Corning, 1946).

20 *United States* v. *General Electric Co.,* 80 F. Supp. 989, 1003 (S.D.N.Y. 1948).

21 *Ibid.,* 1003.

change its attitude, particularly in the light of the pending suit which it alone will have to maintain when the Variable litigation is settled."[22]

In the concrete-block machinery case a district court, in 1949, referred to letters that denote the "objective to accomplish the elimination of almost every competitor through recklessly made threats of suit. . . . It further appears that there were two or three instances where suit was threatened involving machines that agents of defendants had never even seen. The threat of a law suit for infringement should be bona fide and one asserting the exclusive right granted by the United States should not indiscriminately abuse that right. At least an examination should be made of the charged machine and then a law suit if necessary or at least negotiation started."[23]

A manufacturer of plastic artificial eyes, according to a government complaint, coerced others into restraint of trade by threats of infringement suits based on pretended patent rights; a part of the final judgment in 1952 enjoined this alleged practice and ordered the grant of royalty-free licenses under the patent in question.[24]

Threats of infringement suits against customers of rivals were prevalent prior to 1914—the "brigand" era of the patent system. During this era, incidentally, patents were a favorite device of the swindler, as evidenced by the numerous states which imposed police-power regulation on dealers in patent rights. Since 1914, such threats have cropped up now and then. The Federal Trade Commission, organized in 1914, has taken action on this ground against several companies—against National Binding, Nulomoline, Portable Conveying Machinery, American Lecithin, and Pure Carbonic.[25]

It is difficult to prove that threats to bring infringement suits are brought in bad faith. At least two orders of the Federal Trade Commission relating to this practice have been reversed by circuit courts of appeal, in 1925 and 1931—an experience of course which discouraged any further action of this sort.[26]

A recent case, decided December 18, 1952, was a suit by Technical

[22] *United States* v. *General Instrument Corp.*, 87 F. Supp. 157, 172 (D.N.J. 1949).

[23] *United States* v. *Besser Mfg. Co.*, 96 F. Supp. 304, 312 (E.D. Mich. 1951).

[24] *United States* v. *Mager and Gougelman, Inc.*, CCH 1951–52 Trade Cases ¶ 67,233 (N.D. Ill. Civil No. 49–C–1028; Complaint, 1949; Consent Judgment, 1952).

[25] Federal Trade Commission, *Annual Reports* for 1918, 1919, 1941, and 1948; Complaints Nos. 10, 29, 235, 4173, and 5143.

[26] *Heuser* v. *Federal Trade Commission,* 4 F. 2d 632 (7th Cir. 1925); and *Flynn and Emrich Co.* v. *Federal Trade Commission,* 52 F. 2d 836, 838 (4th Cir. 1931).

Tape against Minnesota Mining and Manufacturing "to have the defendant's patent declared invalid, or, if valid, then not infringed."[27] The patent related to adhesive tape sold under the brand name of "Scotch." A district court dismissed the complaint but was reversed by a court of appeals. According to the higher court, Technical Tape applied for a license from Minnesota Mining and Manufacturing and was denied with a warning that if it "made the tape as it contemplated it would face a highly expensive lawsuit."[28] Moreover, Technical was later told that its tape was an infringement and that it would be given the "toughest fight it ever saw."[29] As a result of the litigation threat, Technical's sales "had met with resistance because of its prospective customers fear of action by the defendant" and "it had been forced to give many distributors a written promise that it would undertake the defense of any action for infringement against them."[30] The court held, "Once the patentee has made some claim, directly or indirectly, so that notice is given that it asserts that there is or will be an infringement, a justifiable controversy exists, entitling the alleged infringer to seek declaratory relief."[31]

b. *Malicious Infringement Suits.* Infringement suits may not be brought in good faith. A manufacturer of welding and cutting torches declared in 1936 that his company "after five years of litigation, finally established the invalidity of two of the three patents upon which it had been falsely sued"—patents which had been acquired by Union Carbide and Carbon—and added that patents are too often used "for the purpose of crushing a competitor."[32]

In 1938, a competitor of Automatic Signal Corporation testified that this corporation employed patent suits "to drive out of business all small manufacturers but ourselves, because of their inability to defend such patent cases, although they had their own operating patents."[33] In the same year counsel for a small enterprise stated that a "very large company has started a train of suits, one suit after the other, three suits, and they have declared on certain of the claims

[27] *Technical Tape Corp.* v. *Minnesota Mining and Mfg. Co.*, 200 F. 2nd 876, 876–77 (2d Cir. 1952).
[28] *Ibid.*, 877.
[29] *Ibid.*, 877.
[30] *Ibid.*, 877.
[31] *Ibid.*, 878.
[32] Pooling Hearings, Part 2, 1940.
[33] Comp. L. Hearings, 507.

of the patent before going in; we have had to prepare complete defenses on certain of these claims, and then at the opening of the trial they have relinquished their claims of infringement as to certain claims, and my client in this one particular case spent over $5,000 in preparing only for that one phase of the investigation which was dropped at the trial and for which there can be no recovery under the present law on the part of my client for any of those expenses."[34]

A representative of a glass company testified, in 1938, that it had been continually sued by Hartford-Empire from 1926 to 1938—"A new suit has been filed against us making the same contentions."[35] In reply to the question, "Is the fact that no one has asked you for licenses on your machine, do you think, due to this continued litigation?" he answered that it "would be a deterrent, certainly, to anyone who would attempt to go into the business."[36] Litigation encouraged patent owners, particularly individuals and small companies, to sell to Hartford-Empire. A district court, in condemning this company and its allies, spoke of "litigation expensive beyond the dreams of the average man."[37] In 1932, Hartford-Empire won a suit against Hazel-Atlas for infringing a patent fraudulently obtained. In 1944, after the exposure of its scheme, the Supreme Court ordered the lower court to reverse its judgment against Hazel-Atlas, which in the meantime had become a licensee.[38]

c. *Intentional Infringement*. The large company may defiantly infringe the patents of the independent inventor or his small company and thus in effect deprive him of his legal monopoly. As stated in 1935 by the American Association of Engineers, "Corporations will infringe a valuable invention knowing as it does that the inventor has not the money to fight them in the courts."[39] However, this practice, like malicious infringement suits, is on the decline. To what extent, if any, has this been brought about by exposures of the abuses of our patent system by the TNEC and others?

d. *Exclusive Arrangements*. One may be cut off from the supplies and markets for the manufacture and sale of his invention. According to findings of the Federal Trade Commission, Eastman's percent-

[34] TNEC Hearings, Part 2, 284.
[35] *Ibid.*, 629.
[36] *Ibid.*, 631.
[37] *United States* v. *Hartford-Empire Co.*, 46 F. Supp. 541, 618 (N.D. Ohio, 1942).
[38] *Hazel-Atlas Glass Co.* v. *Hartford-Empire Co.*, 322 U.S. 238 (1944).
[39] Pooling Hearings, Part 2, 1879.

age of the motion-picture film business, because of competition from importers, declined from 94 per cent in 1920 to 81 in 1921. Because Eastman acquired three laboratories for making motion-picture prints and thus threatened to enter this business, manufacturers in this field agreed to use only American-made films so long as it would not compete with them. In 1924, the Commission ordered, first, the discontinuance of this agreement and, second, the sale of these laboratories. The Supreme Court upheld the first order but not the second.[40]

According to a government complaint of 1947, Eastman and Technicolor Motion Pictures Corporation, in 1934, made exclusive agreements with each other covering a process and film. A final judgment of November, 1948, ordered Eastman to grant licenses under certain patents without royalties, and under others with reasonable royalties; to furnish know-how; to refrain from action against any alleged infringer; and to discontinue discrimination in the sale of film. A consent judgment of 1950 required Technicolor to terminate arrangements which forced motion-picture producers and other customers to use only its cameras and other facilities in making color motion pictures and to license its patents.[41]

In discussing GE's monopoly in incandescent lamps a district court said, "Applying its massive dominance over the industry it could by asserting patent control, by controlling and restricting domestic sources of parts and machines, by sealing off the foreign source of parts and machinery through the foreign agreements, and by its exertions in behalf of the Mazda mark, seriously limit domestic competition. All of these conditions generated the power to exclude competitors from the industry."[42]

Exclusive and discriminatory agreements have been made by the government itself. Defense and other departments pay certain corporations for research in their behalf and then let them keep patents on resultant inventions but do nothing of the sort for other companies or independent inventors. Worse still, in contracting for production government departments assume the liability of infringement and then in effect tell the independent inventor to redress his wrongs through ruinous litigation.

40 *Federal Trade Commission* v. *Eastman Kodak Co.*, 274 U.S. 619 (1927).

41 *United States* v. *Technicolor, Inc.*, CCH 1950–51 Trade Cases ¶ 62,586 (S.D. Calif. Civil No. 7507–M; Complaint, 1947; Consent Judgments, 1948 and 1950).

42 *United States* v. *General Electric Co.*, 82 F. Supp. 753, 900 (D.N.J. 1949).

Hired Inventor of Business Corporations

The foregoing difficulties generally do not bother the hired inventor of the business corporation. His employer provides the patent attorney and others for preparing and presenting the application and defending the patent. The company furnishes the capital and know-how for developing and selling the invention. It arranges for any license agreements or sale of patent rights. In contrast with the independent inventor, the hired inventor works under the guidance of superiors and has the advantages of a laboratory, library, and contact with technology. A regular salary frees him from the worries of getting a livelihood. In spite of, and partly because of, these alleged advantages he seldom brings forth a great invention, although he and his fellow-employees are very effective in developing and improving the inventions of others—the telephone, airplane, cellophane, etc. Nylon, a Du Pont product, appears to be the most notable exception. Why is the hired inventor of the business corporation less creative than the independent inventor? The reason lies in a different set of obstacles or handicaps—pecuniary supervision, the goal of monopoly and restraint of trade, and inadequate stimulus.

Pecuniary Supervision. A corporation, like others, does not know how to find potential inventors. It hires scientifically-trained graduates of colleges and overlooks the young Edisons and Fords. Most important, it generally directs its inventors along chosen lines rather than let their imagination have free play. After all, the director of research must make reports of results which point to greater earnings. In other words, money making, rather than delving into the unknown, defines the boundaries of their activities and is the common denominator to which the value of their efforts is reduced sooner or later. Armstrong, De Forest, Farnsworth, and Fessenden—outside the laboratories of American Telephone, RCA, and GE—are among the recent inventors who have made communication what it is today. Suppose their inventive faculties had been under the direction of those companies. How, if at all, would the kind and quality of their inventions have been affected? Someone has said, "A kindred aid is indifference to likelihood of commercial success. It is for these reasons that the lone and independent inventor will remain important in spite of the superior facilities of the organized laboratory."

The great inventor is not confined to present fields of investigation, existing knowledge, and immediate and practical problems of present techniques. It is his role and capacity to find a new frontier in a land already cleared and tilled according to present standards—to discover ways to apply brakes with air as well as physical things, to communicate without as well as with wires, and to fly in heavier as well as lighter than air vehicles.

Monopoly and Restraint of Trade. Pecuniary supervision is especially significant if the corporations which employ inventors monopolize or restrain trade by means of patent pools, consolidations, license agreements, and cartels. They are not interested in bringing forth rival inventions that would displace their products and services or depreciate their existing equipment—a fact which seriously delimits the activity of the hired inventor.

a. *Pools.* The incentive to invent within a pool depends upon the arrangements among the members. If the pool terminates at a definite date or is apt to be disrupted through disagreement or illegality or if any member may withdraw upon short notice, the incentive remains because there still exists the fear of being hemmed in by the inventions of other members. Another important factor in this situation is the likelihood of inventions by outsiders, well illustrated in the contrast between the automobile and airplane pools. Millions of people operate and repair automobiles and thousands sell parts to users as well as manufacturers of them, but comparatively few render the same services for airplanes. Consequently the urge to invent because of competition with outsiders is much greater in the automobile pool than in the airplane.

In recent years the assemblers and manufacturers of parts and the users of those parts account for nearly all the inventions in motor vehicles yet for relatively few of them in airplanes. Stated differently, the government has financed research very little in the one and very much in the other—in 1951, for example, the government paid for about 10 per cent of the research cost in motor vehicles and for about 85 per cent in airplanes.[43]

That a member of a pool may welcome relief from competition in corporate research is indicated in this brief statement, in 1946, by the

[43] *Industrial Research and Development,* U. S. Department of Labor and U. S. Department of Defense *Joint Report* (1953), 11.

Department of Justice: "Patent pooling arrangements authorized during the war in the petroleum, synthetic rubber, and other fields found such favor with elements in the industry fearful of vigorous competition in research that great pressure is brought to bear for the prolongation of these pooling arrangements."[44]

b. *Consolidations.* A patent-owning monopoly seldom brings forth from its laboratories any important development in its field—a fact which seriously delimits the activity of the hired inventor.

The telegraph companies did not invent the cable. Neither the telegraph nor the cable companies invented the telephone. The telegraph, cable, and telephone companies did not invent the wireless telegraph; they even declined to buy it. The telegraph, cable, telephone, and wireless telegraph companies did not invent the wireless telephone. The gas companies did not invent the electric light. The horse-car street railways did not invent the electric railway. Again, steam companies did not invent the steam turbine and the internal-combustion engine. Silk manufacturers did not invent rayon. Collar manufacturers did not invent the soft collar. Manufacturers of washing machines did not invent the continuous laundry. The entrenched shoe-machinery company did not develop the use of glue for attaching soles to shoes. Motion-picture producers did not invent television. Aviation pools and companies, either here or abroad, did not invent the jet engine. The basic concepts in electronics did not come from electric-equipment industries. The manufacturers of dynamite and other mild explosives did not invent the atomic and hydrogen bombs.

Buying up the patents and other assets of rival companies lessens the incentive for research and therefore the opportunity for hired inventors. Former rivals may be required to assign their future patents to the purchaser of their assets without remuneration. Former rivals may even agree to discontinue research for the improvement of their former product.

After the formation of Hartford-Empire, in 1922, and its domination of feeder-type machines for making bottles, an official of Owens-Illinois Company stated "that there is very little to be gained by us in continuing our present effort to conceive new ideas on feeder-type machines."[45] In discussing Hartford-Empire, the Supreme Court

[44] CSB Report, 249.
[45] *United States* v. *Hartford-Empire Co.,* 46 F. Supp. 541, 615 (N.D. Ohio 1942).

spoke of "the offensive and defensive alliance of patent owners with its concomitant stifling of initiative, invention, and competition."[46]

The Parker Rust-Proof Company acquired certain patent rights as well as the physical assets of American Chemical Paint Company. However, according to agreement, "Neither American nor any of its officers was to receive any compensation for improvements on the patents under which Parker had the exclusive license and, therefore, there was no reason why either American or its employees named should continue in their research because they would not be compensated for their time, expenditure or creations made in connection with such investigations."[47]

Preventing the research of former rivals is included in several complaints of the Department of Justice. A paper company, for example, in compliance with an agreement with the A. B. Dick Company, ceased further research toward the perfection of a machine-made tissue for a stencil base.

c. *License Agreements.* A licensor may drive hard bargains with his licensees which have the effect of discouraging their inventors. Chapter Seven has shown the extent to which licensees must accept the validity of the licensor's patents. In many instances the royalties are mere payments for immunity from suits for alleged infringement of invalid patents.

Also licensees of dominant organizations usually must share or surrender their inventions and, as a result, have less or no incentive to invent.

Hartford-Empire leased its machinery to manufacturers of glassware with the requirement that all improvements on the machines become Hartford-Empire's property. These instructions, according to a district court, "further served to destroy initiative in promoting the progress of arts and science."[48]

Licensing by Carboloy was on condition that the licensee or agent should assign or cross-license future patents. In 1948, a district court stated, "The Carboloy company ensured for itself access to any patent rights which any of its manufacturing licensees or agents might subsequently acquire. In the case of the manufacturing licensees,

[46] *Hartford-Empire Co.* v. *United States,* 323 U.S. 386, 407 (1945).
[47] *United States* v. *Parker Rust Proof Co.,* 61 Supp. 805, 811 (E.D. Mich. 1945).
[48] *United States* v. *Hartford-Empire Co.,* 46 F. Supp. 541, 615 (N.D. Ohio, 1942).

Carboloy was either automatically licensed, or was required to be licensed on request, to any such patent rights. In the majority of cases, these licenses, in effect, were exclusive. . . . Both licensees and agents were obligated to do everything possible to secure the rights to any invention made by any of their employees."[49] An inventor and general partner of McKenna Metals Company testified, "Thus it was not to the interest of any of those agents to try to make any inventions. I have talked with many of them about it, and they generally express the notion that it would not pay them to try to make any improvements in cemented carbide or tools because of this agreement which the Carboloy Company made them sign."[50]

General Electric's license agreements with certain other manufacturers required them to cross-license GE under all patents in any way relating to incandescent lamps for the entire life of each patent but their rights to the patents received from GE expired with the licenses. This provision and their quotas of production fixed at a small percentage of GE sales would tend of course to discourage their research and therefore their hired inventors. According to the court, in 1949, the quota system "placed the B licensees in such a position that with their income circumscribed to a given fraction of General Electric sales, they could not support expenditures necessary to operate extended research and engineering development projects."[51] GE also had the "power to set the standard of efficiency of incandescent electric lamps for the entire industry and in so doing to determine what should be their length of life."[52]

GE's fluorescent-lamp licenses were also tied to the incandescent. "In the event the incandescent-lamp license agreement between the parties hereto, dated . . . July 1 . . . 1933, shall be terminated for any reason whatsoever, this present agreement shall terminate simultaneously therewith."[53] The court held that this license provision is unlawful. "The immediate defect of the fluorescent license with the annexed condition is in the implied threat which it held. It had a coercive effect in persuading the licensee to hew to the General Electric incandescent lamp program under penalty of losing fluorescent rights.

[49] *United States* v. *General Electric Co.,* 80 F. Supp 989, 1005 (S.D.N.Y. 1948).
[50] Bone Hearings, Part 4, 1937.
[51] *United States* v. *General Electric Co.,* 82 F. Supp. 753, 858 (D.N.J. 1949).
[52] *Ibid.,* 899.
[53] *United States* v. *General Electric Co.,* 82 F. Supp. 753, 858 (D.N.J. 1949).

In this respect, the restrictive condition brought compliance to General Electric exactions and served as one of the many means to maintain the balance in the industry which it dominated."[54]

Alcoa required its licensees, Reynolds and Kaiser, to let it use their improvements but forbade their exchange of such inventions. According to a district judge, in 1950, "Because an improvement by Reynolds or Kaiser can never injure Alcoa, their major competitor, it appears to me that the smaller companies will likely be discouraged from pursuing, with fullest vigor, technological investigation in the possibilities of the future. This helps to explain, perhaps, why Alcoa's research department had 762 employees in 1949, while Reynolds and Kaiser combined had but about 88. . . . In my opinion it is a vital necessity that there be effective competition among the companies in research and development, as well as in market trade. The consuming public has much to gain from technological improvements, and the security of the country, as well as national prosperity, are largely dependent upon continued scientific advances."[55]

d. *Cartels.* A cartel which divides the field among its members is the worst kind in discouraging inventions. It is illustrated by the agreement which gave control of the petroleum industry to Standard of New Jersey and of the chemical to Farben. Such an arrangement to that extent discouraged chemical research in the United States.

Standard of New Jersey, in keeping with this agreement, licensed its hydrogenation process upon conditions that prevented its use for purposes of the chemical industry. An official of the Texas Company protested as follows: "It seems economically wrong to restrict a licensee from selling any product which is a direct result of the hydrogenation process under which he is licensed."[56] And, in reply to the question, "You never made rubber prior to 1939 in your own laboratories; you were always depending upon somebody else prior to that time?" a chemical engineer for Standard Oil Development said, "That would be my recollection."[57] In order to operate the hydrogenation process which Standard of New Jersey controlled, licensees had to buy their catalysts from its subsidiary. One of them, the Texas Com-

[54] *Ibid.*, 858.

[55] *United States* v. *Aluminum Co. of America,* 91 F. Supp. 333, 410 (S.D.N.Y. 1950).

[56] Bone Hearings, Part 7, 3345.

[57] *Ibid.*, Part 9, 5170.

pany, declared that this arrangement "seems to restrict development rather than encourage and promote it."[58]

Lack of Stimulus. Inventing in big corporations is as highly specialized as manufacturing, transporting, record keeping, and other phases of business. Corporate invention is like an assembly line along which each worker does a part. The personality of the hired inventor cannot be incorporated into the invention any more than that of the laborer can be in the final product. Perhaps one makes a temperature test and the other adds a bolt along the assembly line. Such cog-like work leaves little opportunity for the instinct of workmanship. Its deadening effect is bound to clash with the prerequisites of invention —initiative, originality, and imagination.

Can a great invention emerge at the end of the assembly line? Can the great invention be created by many, each adding something here or there, each a specialist in his own field, all working together as a team for a certain effect? Can only a genius conceive or create the great invention?

According to Nikola Tesla, "Invention is predominantly individualistic. Everything of prime moment comes from some individual unconnected with any commercial organization—from the lone worker who follows the fleeting inspiration of a moment and finally does something that has not been done before."[59]

The hired inventor receives a salary as reward rather than a patent with all its possibilities. He escapes the trials and tribulations of the independent inventor but foregoes that remote chance of reward which may come from the exclusive right to make, use, and sell. He makes a sure living by digging for gold for others, rather than by staking a claim and digging for gold for himself. Yet the lure of the big reward has influence among inventors, especially the mine-run variety. If this is not true, the forefathers of the nation proceeded under a false assumption in authorizing Congress "to promote the progress of science and the useful arts by securing for limited times to authors and inventors, the exclusive rights to their respective writings and discoveries."

At one time a laborer retained all the patent rights to his invention except a so-called shop right, or license to use it, of his employer;

[58] *Ibid.*, Part 7, 3342.
[59] Pooling Hearings, Part 1, 686.

but today in some companies, the laborer, like the hired inventor, is compelled to assign his patent to the employer. Neither one has even the delusion arising from patents to stimulate his effort.

According to Frank B. Jewett, Bell Laboratories at first paid something extra to an employee in addition to salary if he took out a patent and assigned it to the company. Jewett said that this system of remuneration put a tremendous incentive on the employee to work for himself at the expense of his employer and, worse than that, put him in the position of working in competition with his own associates.

> The incentive was to get out as many patents that would pass the Patent Office as possible. An invention was made. It could be covered by one strong patent or it could be covered by a dozen minor patents. It was to the company's advantage to have one strong patent, but it was to the employee's advantage to have a dozen minor patents, because he profited in a monetary sense from that.
>
> It created a situation where men would not work with each other, they would not confide with each other, yet the problem which was before us was a problem which required team action; because these developments in the field of communication, as in all these other modern industrial fields, are too vast for any individual to cover completely. So some way had to be found to get over that.
>
> Then, in addition to that, it is only a small fraction of the things which are done in research and development laboratory that come within the purview of the patent law. It is only those things which are new and novel, and which have not been practiced before, which come within the things which the law says can be patented.
>
> We had the situation where two-thirds of our people, may be three-fourths of our people, were doing work which under no conceivable condition could result in patents, who were being remunerated wholly through salary, and right alongside of them were these people who were getting both salary and something else.[60]

Other large corporations, perhaps with an experience like that of Bell Laboratories, generally give no special awards to hired inventors. According to International Harvester, in 1935, employed inventors "are under contract to assign their inventions relating to the company's product without other compensation than is provided for in their employment agreements."[61] Edsel Ford testified, in 1938, "Em-

[60] *Ibid.*, 276.
[61] Pooling Hearings, Part 4, 3413.

ployees that create inventions of their own origin and are made on company's time are assigned to the Ford Motor Company. We pay them no fee for that invention."[62] According to Charles F. Kettering, vice-president of General Motors, "If you gave the reward to a particular individual for his particular invention, then he would be secretive about the thing, so we try to reward the whole laboratory."[63]

University and Government Researchers

The university professor and his counterpart in government service may have the opportunties of both independent and captive inventors and yet may avoid their handicaps. After performing their regular duties, they may have the time and energy and facilities for delving deeper in their own specialty or following paths into new frontiers. If fortunate enough to be relieved from daily routine, they are able to dedicate themselves completely to the quest. The library and laboratory nearby may reveal what is and lead the way to what is yet to be. Their academic training gives them scientific knowledge and method and yet perhaps leaves them free to doubt and explore. They are not under close and practical supervision with profit-making as an incentive. Their salaries may eliminate the worries about food, clothing, and shelter and therefore even about patents. Some of our greatest inventions, as illustrated already, have come from the university and government source, and the expense has been nominal—results that compare extremely favorably with those from the research of business corporations. Of course, the research of professors, if financed by a private corporation, may suffer from the same handicaps and for the same reasons that research in the corporation's own laboratory suffers. The captive contributes no more or less in one place than another.

Conclusion

Discouragement of the independent inventor tends to lessen invention. The assumption underlying the grant of patents is a legal monopoly or possible reward to the inventor in return for the disclosure of his invention. His chance today of receiving enough to pay expenses of

[62] TNEC Hearings, Part 2, 262.
[63] *Ibid.*, 341.

getting a patent and then defending it is very meager. His stimulation comes through delusion rather than reward. But the chance of a lucky strike, however remote, exists, and hope springs eternal. The hired inventor, though assured of a livelihood, suffers from pecuniary supervision and lack of stimulus. Licensees that assign their inventions ordinarily have neither delusion nor remuneration. If individuals invent as much as ever, it is for reasons besides a big reward. All things considered, university and government researchers, especially the former, seem to have the most favorable environment for inventing. Is there still another and better way of encouraging the inventor to give us the benefit of his genius?

CHAPTER X

Remedies

THE PATENT SYSTEM of the United States, since 1836, has reflected changes in economic environment rather than changes in patent legislation. The system has responded to industrial development, to the various kinds and increased number of inventions, to the new method of invention brought about by the emergence of hired inventors, and to the limitations put on the use of inventions by antitrust laws and the judicial interpretations of those laws. These changes in the economic environment which supports invention have influenced the patent system but have not actually contributed to any basic changes in the patent laws themselves; therefore the patent system has been able only partially to adapt itself to the demands of the economic world in which we now live. The demands of an economic world which is the result of a century of economic growth require us to appraise anew the effect of patents, the need for them, and proposals for improvement in the patent system.

Basic Changes in Patent System

Stage of Industrial Development. The effect of a patent system upon the economic condition of a country depends upon the industrial development the country has achieved. Germany, for example, refused to grant patents until 1877, until it caught up with technological developments in other nations; since then and until World War II, Germany not only granted patents but subsidized the research and inventions which preceded them. Such a program was well suited to them, delayed as they were in their industrial development. Other countries, especially in more recent times, have adopted political and economic systems which in effect permit the use of foreign inventions without payment.

Before as well as after the Industrial Revolution, two common schemes for encouraging new industries were patents and a tariff. In the United States both of them, along with an abundance of natural resources, a large market, and scarcity of labor, hastened the invention and acceptance of power-driven machines. But within recent decades such industries as the chemical, electrical, and aluminum have been organized and developed to such a point that they are no longer dependent upon patents for inventions or a tariff for a market. In both respects they have long since emerged from the infant stage. However, both patents and the tariff have been handmaids in such industries in monopolizing and restraining trade both here and abroad. Any general benefit from either patents or a tariff now lies primarily in new fields where individuals and small enterprises are trying to get under way with new ventures.

Kind and Number of Inventions. The level of technological development which a nation has achieved determines, at any given time, the kind and number of inventions the nation produces. For several decades, after the beginning of the United States patent system, practically all the inventions came from artisans and the handicrafts and, therefore, were simple and recognizable. The Industrial Revolution in this country did not get well under way until about 1850. Since then, inventions to an increasing extent have been in scientific and complicated fields, at first in power-driven machinery especially and later in electricity and chemistry. Instead of a few improvements on a jackknife we now have thousands on the telephone and electric lamp. Millions of patents issued by this and other governments, with thousands of patents on endless variations and improvements of one basic invention, result in overlapping patents and therefore in the conflict and invalidity of many of them.

Associated with the change in the kind of invention is the extent of incomplete disclosure of the invention in the patent and, therefore, the importance of know-how or secret technology. The omission of key steps in certain inventions such as synthetic rubber cannot be readily detected by examiners. Moreover, the nebulous state of the invention or discovery may prevent exact descriptions. Whatever the explanation, the result is pre-emption, by indefiniteness, of certain frontiers of technology and therefore the defect of both the patent and antitrust laws.

Overlapping patents and incomplete disclosure have become worse because of the increasing inability of the Patent Office to examine applications as effectively as they are presented.

Rise of Hired Inventors. A significant result of the Industrial Revolution which must be kept in mind in any remedial program for patents is the decline in importance of the independent inventor and the rise of the hired inventor. The independent inventor roams the field and finds his "reward" in the grant of patents; he still brings forth the best and the worst of inventions though under bigger handicaps than in the past. In contrast, the hired inventor is another employee on a salary and the grant of a patent on an invention in which he participated has little or no causal relation to his efforts. The expense of such research—salaries, laboratories, etc.—for the going concern is like the expense of time and motion studies, market research, and the like in that it is necessary in order to keep up or stay ahead. In a competitive setup the reward of corporate research is the advantage, however temporary, a company has in taking advantage of its findings. In all instances it must conceive, develop, and reduce ideas to practice, whether they be improvements in machinery, assembly line, or sales technique; patents apparently play little or no part as an incentive. The grant of patents to a corporation via its employees is another inane effort to adapt the legal theory of a corporate person to a law originally intended for the individual. The corporation has largely usurped the patent system as it once did the fourteenth amendment. Insofar as the big corporation and the hired inventor, one receiving the patents and the other wages or salary, has replaced the independent inventor, the "promotion of the progress of science and the useful arts by securing for limited times to inventors, the exclusive rights to their respective discoveries" has ceased.

Antitrust laws. Any patent system is influenced by the absence or presence of antitrust laws. In this country, patents have been manipulated in sundry ways in seeking immunity from the Sherman Act and kindred legislation. Pools and consolidations are primarily means for bringing together the patents of an industry in order to control it. License agreements provide the facility by which former and prospective competitors may become licensees under restrictions as to output and price. Cartels, in effect, are international pools, consolidations, and license agreements. The dictation of supplementary supplies widens the monopoly. Tests of the validity of patents are generally avoided because of expense and uncertainty of outcome. Basing agreements on a most recent patent is one of several schemes to prolong a monopoly. Also patents have been employed, though to a decreasing extent, in competing unfairly and thus holding would-be rivals at bay or bringing them to terms. Many patents are suppressed, and invention is discouraged.

In other words, dominance in an industry has been achieved by bringing relevant patents under a single control and thus pre-empting the field. This has been extended in various directions through license agreements. A false but defiant front is maintained by avoiding the tests of validity. Most of the purported patent strength is deadwood, but still alive in appearance. It is continued indefinitely by taking out more and more patents. This jungle of briars and thorns permits the strong to exclude or entangle the weak. Alternate inventions thus patented and held seldom emerge as products for daily use. The independent inventor, like the would-be competitor, cannot enter to buy or sell a patent right except upon dictated terms. Under this condition invention becomes primarily a means for establishing and maintaining industrial monopoly and restraint of trade.

Judicial Interpretation. When does a patentee merely exercise the right, as intended, of excluding others from making, using, and selling his invention? When does he use such a right in violation of antitrust legislation? These questions have come before the courts in many cases; as they are answered one way or another the patent system changes. Witness the judicial interpretation of the control of supplementary supplies by the patentee and the effect of the interpretation in the expansion or contraction of his practices. Congress adopts laws and the courts must interpret them, particularly when the laws clash with each other.

Methods of Improvement

Methods for overcoming the abuses of our patent system are to be found in private enterprise and in the government. Methods of the government include patent and antitrust legislation, administration by the Patent Office, and interpretation by the courts.

Private Enterprise. Any enterprise charged with infringement may challenge the validity of the patent in question and, if successful, escape the payment of damages. Another defense is to demonstrate the misuse of the patent by the owner. As a rule, however, a company charged with infringement cannot afford such remedies because of the expense of patent litigation. A more effective strategy—this also is reserved for only a few—is to match any threat from a patent owner with a threat to withdraw patronage or to become a competitor. Ford Motor, for instance, defended itself both by legal action and by threat. Ford defied the Selden and other patents, forced the issue of invalidity, and won. Ford's practice has been to collect no royalties on its own patents, and with relatively few exceptions it has paid none on the patents of others. Also, an employee of Ford indicated "that they were doing a going rate of ten or twelve million dollars a year with General Electric and so should not have any trouble about patents" on tungsten.[1] In addition, Ford and General Motors made or threatened to make their own tungsten and as a result got a reduction in price from $205 to $45 per pound. According to a district court, they "made or threatened to make their own material in order to get lower prices," and "the record indicates that the real purpose" of the threatened suits against them was to sell material to them rather than to sue them.[2]

It should be added that in a few instances the most effective check upon industrial monopoly has resulted from the head-on clash of giants. Swift and Armour packers and International Shoe clipped the leather trust by going into the tanning business. International Shoe obtained certain concessions from United Shoe Machinery by threatening to build its own shoe machinery. Again Ford Motor and other concerns, by making their own steel, were largely responsible for disrupting the "Pittsburgh plus" scheme for pricing this metal. Also, manufacturers of automobiles insist upon competition in parts. They avoid dependence if possible upon one source. As one manufacturer of wheels said, "We realize also that all of the large manufacturers of automobiles are not going to confine themselves to one source of supply."[3] Unfortunately, no one user of such things as incandescent lamps, cameras, glass containers, mimeographs, cash registers, and harvesting machines purchases enough of them to justify the develop-

1 Bone Hearings, Part 1, 149.
2 80 Fed. Supp. 989, 1003 (S.D.N.Y. 1948).
3 TNEC Hearings, Part 3, 1058.

ment of competing improvements and the manufacture of his own supply.

Legislation. There has been no legislation designed primarily to remedy abuses of our patent system. The nearest approach to laws for this purpose is found in the Clayton Act. Section 3 of that act reflects opposition to the dictation of supplementary products for a patented invention and to the approval of the practice by the Supreme Court in the mimeograph case. The act specifically forbids the sale or lease of products, whether patented or unpatented, on the condition that the buyer or lessee shall not use or deal in the goods of a competitor if the effect "may be to substantially lessen competition or tend to create a monopoly in any line of commerce." The law states "whether patented or unpatented," and therefore definitely stipulates in effect that patented goods provide no exception. As already explained in Chapter Six, an effort to offset this law and modify the mimeograph decision in the projecting machine and subsequent cases is the new definition of contributory infringement in the Patent Act of 1952. Will the courts ignore this definition as they apply the Clayton provision in the future?

In 1950, Congress, in amending Section 7 of the Clayton Act, forbade one corporation to acquire the assets of another if the effect may be substantially to lessen competition or tend to create a monopoly. It is expected that this provision will apply to patents as well as physical assets.

This dearth of legislation for improving our patent system cannot be attributed to the lack of information or recommendations from government agencies and others. Footnotes in previous chapters refer to congressional hearings and reports relating to patent abuses and possible remedies—Oldfield (House), 1912; Nolan (Senate), 1919; Pooling of Patents (House), 1935; Compulsory Licensing of Patents (House), 1938; and Bone (Senate), 1942. Other former patents committees of both the House and Senate have had before them numerous bills, some of which would extend and others restrict the rights of patentees. Similar committees have included patents in their surveys; for example, Truman (Senate), 1941–1945; Elimination of German Resources for War (Senate), 1945; United States versus Economic Concentration and Monopoly (House Committee on Small Business), 1946; and Cellar (House) beginning in 1949. The TNEC, made up of representatives of Congress and departments and commissions, made an exhaustive report entitled "Investigation of Concentration of Economic Power." It approved eight proposals by the Department of Justice for changes in the patent system. Various commissions have issued reports involving patents; especially noteworthy is the FCC report of 1939 on the telephone industry. The Department of Justice and the Department of Commerce have sponsored studies; worthy of particular notice is the former's Investigation of Government Patent Practices and Policies. The National Patent Planning Commission, appointed by President Franklin D. Roosevelt, issued two reports, in 1943 and 1944. In addition there have been numerous books and pamphlets by students of the patent problem and sundry reports by patent bar associations and other groups, some of which are included in the bibliography of this book.

Administration by Patent Office. A commissioner of patents may adopt rules of practice which discourage certain abuses. Thomas Ewing, commissioner of patents from 1913 to 1918, said in 1916, "To count on Congress for large improvement in conditions seems to me, in the light of experience, to be mere folly."[4] Consequently he effected some reforms of his own; for example, pressure on any application pending over three years. In response to his efforts the court of appeals decided that, in case of dilatory prosecution, the ordinary rule that all doubts as to patentability be resolved in favor of applicant should be reversed.[5]

Judicial Interpretation. After the adoption of the Sherman Act the first decisions of the federal courts indicated the exemption of the patentee from its provisions. In several instances they upheld pools, consolidations, license agreements, and control of supplementary goods in spite of this law. But there were unfavorable decisions in the most flagrant instances, illustrated by the bathtub case of 1912.[6] The first clear-cut reversals came in the Sanatogen[7] and projecting-machine[8] cases, in 1913 and 1917 respectively, in which the Supreme Court reversed its position of 1912. Restrictive agreements imposed upon former competitors through license agreements as to price and

4 *Ibid.*, Part 2, 723.
5 *Ibid.*, 723.
6 226 U.S. 20 (1912).
7 229 U.S. 1 (1913).
8 243 U.S. 502 (1917).

output which were accepted by the Supreme Court in the Bement case of 1902[9] were declared illegal in the Gypsum[10] and Line Material[11] cases of 1948. The sanction of control of prices of licensees in the General Electric decision of 1926[12] was thrown into doubt by a four to four division concerning it in the Line Material decision just mentioned. The Supreme Court did not have an opportunity to pass upon a cartel until the National Lead case in 1947.[13] As various schemes have been declared illegal, variations of them have been tried over and over again.

More recent decisions, whether involving patents or not, have shown a greater willingness to apply the Sherman Act and other antitrust legislation. A notable exception has been the decision concerning consolidation of physical assets or patents. The Supreme Court has never condemned this practice as it has restraint of trade. The court has taken no stand against the merger of the physical property or patents of two or more companies, but many times it has disapproved restraint of trade between or among them. This has encouraged the consolidation of former competitors until one or a few of them become dominant in their industry—until they become a monopoly or an oligopoly.

Recent decisions of the Supreme Court relating to validity of patents and infringement have weakened the legal position of those who use patents to defeat antitrust legislation. In the Sola Electric decision of 1947,[14] the Supreme Court held that the validity of a patent may be challenged by a licensee if his contract, unsupported by valid patents, would be illegal. This decision gives an additional incentive to achieve control of an industry through consolidation of assets rather than license agreements. Another blow against the effort to conceal invalidity came in 1948, when the Supreme Court declared in the Gypsum case[15] that the Department of Justice may attack the validity of a patent in an antitrust proceeding. And in the Leitch case of 1938,[16] this tribunal held that patents used in violation of the Sherman Act

9 186 U.S. 70 (1902).
10 333 U.S. 364 (1948).
11 333 U.S. 287 (1948).
12 272 U.S. 476 (1926).
13 332 U.S. 319 (1947).
14 317 U.S. 173 (1947).
15 333 U.S. 364 (1948).
16 302 U.S. 458 (1938).

may be denied relief against infringement until the effects of this misuse are removed. Consequently, if a patent-owning corporation observes the antitrust laws its patents are less likely to be tested for validity and to be infringed by others. On the other hand, if it violates them it runs a great risk of having its patents either tested and declared invalid or infringed with impunity. In effect, patents lose their legal effectiveness while being used in violation of antitrust laws.

The Supreme Court, in Hartford-Empire[17] and National Lead[18] decisions, held that the Department of Justice in an antitrust case can secure compulsory licensing of patents at a reasonable royalty for both present and future patents. In Hartford-Empire the court required "reasonable royalties" on patents and thus modified an antitrust decree of a lower court which had ordered licenses without charge to all applicants.[19] In National Lead, it upheld reasonable royalties decreed by a district court.[20] A district court, in 1953, ordered that certain patents on incandescent lamps be dedicated to the public. Recent consent decrees involving the Dick[21] and other companies required, as previously stated, the licensing of patents without royalties and the revelation of secret processes and know-how.

In interpreting invention and patentability the Supreme Court has stressed both "flash of genius" and "innovation"; for example, flash of genius was stressed in the Cuno case of 1941[22] and innovation in the Sinclair and Carroll decision of 1945.[23] Both approaches lead apparently to the same prerequisite for a patentable invention—a contribution over and above the exercise of mechanical skill.

Underlying Objectives

In attempting to bring our patent system into closer harmony with the environment of today it is well to keep in mind two basic objectives. One of them is invention and the other is competition.

Encouragement of Invention. One of the ways to encourage inven-

17 323 U.S. 386 (1945); and 324 U.S. 570 (1945).
18 332 U.S. 319 (1947).
19 46 F. Supp. 541 (1942).
20 63 F. Supp. 513 (1945); and 332 U.S. 319 (1947).
21 *United States* v. *A. B. Dick Co.*, CCH 1948–49 Trade Cases ¶ 62,233 (N.D. Ohio Civil No. 24188; Complaint, 1946; Consent Judgment, 1948).
22 314 U.S. 84 (1941).
23 325 U.S. 327 (1945).

tion is to give a reward to the inventor; this reward may be in money and public recognition as in Napoleon's regime. Or encouragement may consist of securing to inventors for a limited time the right to exclude others from making, using, and selling their respective discoveries; any reward comes from these exclusive rights. Another way is that used in assembly-line research where workers receive wages and salaries; such research, like mass fabrication and sales, is a group activity and the cost is a part of current expenses, and, according to the testimony of business executives already noted, research-invention would continue without patents. Likewise the professors in our universities, often initiating and completing research projects of their own, have sufficient incentive as evidenced by their willingness to give their discoveries and inventions to the public. Any change in our patent system, therefore, should draw a sharp distinction between the independent inventor who may respond to prospective reward from a patent on the one hand and the hired inventor or technician on the other, who, like the time and motion expert or sales strategist, does not require it. In other words, any patent system in the future, like the one in the beginning, should be adapted primarily to the individual inventor.

People learn of inventions and satisfy wants with them through use rather than description alone. Mere description and nothing else is not the equivalent of description and practical application. An invention suppressed during the life of the patent covering it is seldom used after the patent expires because of obsolescence; the most common effect of patent protection of a suppressed invention is to obstruct others for seventeen years. Worse still is an invention incompletely disclosed—one which cannot be put into practice because the know-how is not revealed. A patent on it, as illustrated already in the case of synthetic rubber and other chemical processes, is, temporarily at least, a legalized monopoly from the public with nothing given in return.

Any change in our patent system should differentiate between independent and hired inventors and between inventions used and revealed and those suppressed and hidden. Such distinctions embodied in laws might be challenged on legal grounds, but they would have the merit of logic in the light of the purpose of our patent system.

Competition. A second objective of a patent system should be competition among rival and alternate inventions. There is nothing inconsistent between this ideal and a patent upon each specific invention. In fact, under normal conditions competition is stimulated rather than prevented by patents. The owner of a patent on a new product or a variation of an old one has protection for something new and different—the same advantage that a trade-mark or a tariff may give —and therefore is in a better position to hold his own. This provides an incentive for conceiving an invention and for seeking an opportunity to introduce it.

There are electric and other kinds of refrigeration and variations of each, all patented; competition in price and quality exists. Automobile manufacturers have rival transmission systems and variations of each. On the other hand there are different methods of attaching soles to shoes and improvements of each, some of which were originally suppressed through patent control by United Shoe Machinery. Also, inventors brought forth different kinds of communication—telegraph, telephone, radio, etc.—and variations of each, but they have not been highly competitive because American Telephone entered into a division-of-fields agreement with GE and RCA.

The concentration of patents in closed pools or consolidations, the greatest obstacle to would-be competitors, leads to the suppression of inventions which should be in competition with each other. Analogies may emphasize the gravity of this situation. Suppose one publisher acquired all copyrights on books in a certain field and refused to publish most of them; or that one railway company, obtaining the franchises on the different routes between two points, say New York and Chicago, used only one of them and kept other railways from using the others. Would such actions be tolerated? Likewise license agreements unduly restrictive may produce a result similar to that of monopoly. The Supreme Court recently said, "The stimulus to seek competitive inventions is reduced by the mutually advantageous price-fixing arrangement."[24]

The government may seek relief in general from such situations as monopoly by enforcing antitrust legislation, by regulating prices as it does the rates of railways, by providing government competition as illustrated by TVA and parcel post, or by its ownership and operation of the business. Domination of an industry, however achieved,

[24] *United States* v. *Line Material Co.*, 333 U.S. 287, 311 (1948).

means big business and this in turn, as recent experience shows, may lead to big labor, and both—big business and big labor—may in turn result in big government. As totalitarianism in any of these fields increases, a democracy of individualism and competition must give way. Our ideal is competition, with limits upon acquisitive practices as provided in such laws as the Clayton, Federal Trade Commission, and Robinson-Patman acts. Competition is better for business because it results in a smaller scale of bargaining by labor and in less government activity; most important, competition brings about better products and lower prices for the public.

Better Patents

The weakness of patents themselves—over one-half of the adjudicated ones are declared invalid—indicates the logic of diagnosing applications in terms of a standard of patentability, the need of adequate records and sufficient personnel in the Patent Office, and the need of an increasingly critical attitude on the part of the Patent Office toward patent applications.

Standard of Patentability. Whether emphasis be on "the quality of the inventor's mind" or the "quality of his product" in determining patentability, the imperative need is a higher standard. The Supreme Court has pointed the way to such a standard and the Patent Office and lower courts should follow. The lower courts may, if they deem it necessary, obtain helpful information from physicists, chemists, and the like—an arrangement more satisfactory than a court of patent appeals or any other special technical tribunal with final adjudication as to issues of validity. Any standard of course has its gradations and problems of administration, but the standard, whatever it is, should be observed strictly and consistently. As District Judge William C. Clark of New Jersey once testified, "The interest of the public is more injured by a quantity of unsound patents than allowing one good patent to be rejected."[25] Can a high standard be administered and still have hundreds of improvements piled together like accretions in a coral island, no one being able to tell exactly where one begins and another ends? A higher standard would mean the denial of patents on inventions which cannot be clearly defined and are

[25] Pooling Hearings, Part 1, 1084.

void of any demonstrable importance. Every claim of a patent should be more fundamentally differentiated from the claims of other patents.

The Patent Act of 1952, instead of recognizing the extent to which invalid patents have been issued and of seeking to raise the standard of invention, merely reaffirms that "a patent shall be presumed valid." Is this a sound basis for expecting a decline in the percentage of adjudicated patents held invalid by the courts?

Worthy of appraisal at this point are the so-called opposition proceedings: the Patent Office publishes an application after its acceptance and then permits others to oppose it as unworthy of a patent grant. This idea seems meritorious in that it presumably would provide a check upon the work of examiners and thus prevent the issue of invalid patents. Such a proposal might work as intended in a competitive system. However, if a few large companies control the field, would any one of them challenge the applications of the others any more than it now challenges their patents? Such a question is even more pertinent if these companies pool their patents or cross-license each other. Also, if the field in which the invention lies is dominated by one concern, there would be little or no occasion for a challenge. And if a dominating concern has license agreements with others based on the concern's future as well as present patents, the licensees ordinarily would have little incentive to challenge the applications for those patents. In all these situations, however, companies already entrenched behind patents would have the incentive as well as the finances to challenge would-be intruders. On the other hand, opposition proceedings and therefore the necessity of supporting an application by the lone inventor might place upon him a burden as heavy as the one he now bears in defending his patent. Cost of litigation, already high, would be increased still further, thus putting to still greater disadvantage the poor and small as compared to the rich and big. Burdening the lone inventor and the lack of competition are the main objections to opposition proceedings. It is better for a patent office to perform the entire examination at government expense and thus subject all patents to the same scrutiny.

Records and Personnel. Obvious essentials in examining applications for patents are sufficient records and efficient personnel. The Patent Office should have copies of patents issued by other nations and of pertinent books, magazines, and other sources of information for determining whether an alleged invention qualifies for a patent. Also, the Patent Office should have at least a few physicists and chemists who can readily read foreign languages of importance. The ideal, though it cannot be completely attained, is to have a personnel whose training and experience are a match for the attorneys, engineers, and technicians of corporations who seek patents. If the clash of wits is not upon equal terms, the standard of patentability is bound to give way. According to the Patent Office, a recent study concerning resignations indicates that the average loss in patent examiners is 8 per cent annually.

Attitude. Even more important in improving the patent system is the attitude of those who administer it. There should be the same opportunity and vigilance in refusing patents as in granting them. The alleged inventor has no inherent right to a monopoly. The patent system is for the public. The main difficulty in attaining this ideal is the inherent weakness in an administrative process in behalf of one party only.

If and when the Patent Office administers the standard of patentability indicated by the Supreme Court, the number of patents should be reduced at least one-half. The granting of fewer patents would in turn lead to fewer applications and the need for fewer examiners; moreover, it would reduce correspondingly the need of taking out so-called defensive patents. As the vice-president of Wm. Wrigley, Jr. Co. stated: "If the Patent Office drew a tighter line distinguishing between ordinary mechanical skill and invention, it would probably be unnecessary for us to waste our money in taking out patents which we and our attorneys feel are merely defensive instruments having little, if any, merit, but which we are obliged to attempt to secure patent protection on for fear that some machinery manufacturer might secure limited protection and subject us to a law suit which would cost us more to defend than it does for us to take out innumerable patents."[26]

Elimination of Invalid and Worthless Patents

The total number of unexpired patents in 1954 was around six hundred thousand. A small percentage of them covered inventions em-

[26] Pooling Hearings, Part 4, 3582.

bodied in products on the market, some were suppressed, but most of them were unworthy of development. Whether the number of patents that are invalid is approximately no more than the number of adjudicated patents is debatable. In any event the courts should use every opportunity to expose invalid patents. According to the Supreme Court, "There has been a tendency among the lower federal courts in infringement suits to dispose of them where possible on the ground of noninfringement without going into the question of validity of the patent. It has come to be recognized, however, that of the two questions, validity has the greater public importance."[27] Invalid and worthless and suppressed patents are significant only as obstacles. The obvious thing to do is to get rid of them. One way is to cancel or suspend them for being invalid or in violation of law, and another is to eliminate them through taxation.

Cancellation or Suspension. Surely no one questions the need or justice of exposing invalid patents. In the Sola Electric decision[28] the Supreme Court recognized the right of a licensee to challenge the validity of the licensor's patent—in spite of his agreement to the contrary—if he enters into an agreement which would be unlawful without the support of valid patents. To deny such a right of challenge on the part of a licensee would be practically to prohibit any challenging of patents in those industries in which all competitors are licensed under such a non-challenging restriction. In addition, the Department of Justice should have the right to challenge the validity of any patent which is involved in any antitrust suit. The application of the clean-hands doctrine, in effect, suspends patents while used in violation of antitrust laws.

Annual Fees. If an invention is inherently weak or worthless, the public is better off in letting it lie dormant than in having it put into practice. Nevertheless, the patent covering it stands in the way of the making and using of other inventions in the field to which the patent belongs. An effective and desirable method to eliminate such a patent is the imposition of annual progressive fees or taxes upon the ownership or exclusive license of a patent, beginning some three or four years after the patent's issue. Within the three or four years the invention may become remunerative. However, this fee should be

27 *Sinclair and Carroll Co.* v. *Interchemical Corp.*, 325 U.S. 327, 330 (1945).
28 317 U.S. 173 (1942).

lower on the patents the inventor owns than on the patents which have been acquired from others. In fact, it would be well to exempt the inventor for one-half the life of his patent. This arrangement would enable him to develop his invention if worthwhile or to find a buyer for it before the first annual fee became due.

England affords an excellent example of what may be accomplished by such a tax. The maximum annual fees are £10 the fifth, sixth, seventh, and eighth years, £ 15 the ninth and tenth years, and £ 20 the eleventh to the sixteenth inclusive. Of the patents "sealed" or granted in 1933, 45.6 per cent had lapsed in 1939, 76.4 per cent by 1944, and 85.5 per cent by 1948.[29]

Such a tax, in addition to clearing away worthless patents, would discourage acquiring and keeping patents to dominate an industry. Also, it would be an additional burden upon the suppression of patents. Even more important, it would encourage invention by removing obstructive patents. Thousands of such obstructive patents, taken out both here and abroad, now remain in effect here but are surrendered abroad because this country imposes no annual tax on patents while foreign countries do. Here we permit patent debris to accumulate; elsewhere it is taxed away.

Less Litigation

The independent inventor still finds that his greatest obstacle consists of interference proceedings in the Patent Office and litigation in the courts. He finds it necessary to sue and be sued in order to protect and assert his rights. As shown in Chapter Seven, litigation involving the validity of corporation-owned patents has declined in certain fields. In some instances, the companies in the same industry have agreed, in effect, not to sue each other, or one company in an industry may buy the patents of others and thus prevent any possible clash; another explanation is that with the increasing percentage of adjudicated patents being held invalid, corporations avoid any litigation if at all possible.

In administering the patent system, the procedure should be simplified and appeals reduced so that the independent inventor can afford

29 Floyd L. Vaughan, "Important Differences in U. S. and U. K. Patent Systems," *Journal of the Patent Office Society,* Vol. XXXIII (Nov., 1951), 779–99.

as well as obtain justice; as previously stated, there should be vigilance in protecting alike the rights both of the public and the inventor. Worthy of serious consideration is the use of administrative procedure for infringement suits—a procedure illustrated today by labor boards and workmen's compensation or industrial commissions.[30] Such a procedure would not become a "captive" because there would be two litigants opposing each other. Administrative procedure, as shown by experience in England, is particularly well adapted to determining proper awards by the government to inventors for the use of their inventions.

Antitrust Proposals

As already noted, an alleged infringer may employ the clean-hands doctrine as a defense, and a licensee may question the validity of a licensor's patent if he enters into an agreement with the licensor which would be illegal if the patent is invalid. Both rights should discourage various schemes to monopolize and restrain trade. Both rights would have additional force if the Department of Justice were permitted to become a party to an interference proceeding or an infringement suit when the department's participation were deemed necessary to protect the public interest. In addition to these two means—clean-hands doctrine and the challenging of patent validity—there are other means of improvement which apply only or mainly to a particular scheme for using patents in violation of antitrust laws.

Pools. Pooling or cross-licensing patents, though necessary in certain instances to avoid conflict of technology, may also restrain competition, discourage invention, and uphold invalid patents.

Any thorough appraisal of pools must allow for these possibilities. In the gasoline-cracking case,[31] for example, the Supreme Court, in interpreting the effect of the patent pool upon competition, observed that the interchange of patent rights is frequently necessary if technical advancement is not to be blocked by litigation.

Former competitors that pool their patents with restrictions on price or output obviously restrain trade. If the pool is on a closed-union basis, the members may prevent outside competition by with-

[30] See John C. Stedman, "Invention and Public Policy," *Law and Contemporary Problems,* Vol. XII (Autumn, 1947), 649–79.

[31] 283 U.S. 163, 171 (1931).

holding important technology. This fact emphasizes the importance of admitting others to the pool on terms which encourage them to compete.

A pool of all patents, present and future, discourages invention among members because the inventions of each become available to all. Consequently there is less or no incentive in creating and developing alternate inventions and rival products and processes in the future—a high price to pay for the removal of clash and litigation in the present. A member has less interest even in taking out a patent on the invention of an employee if it must be shared with others. Moreover, a pool discourages invention by the independent inventor because there is less or no competition in purchasing his patents.

The arrangement made for nonexclusive licenses among refiners of cracked gasoline brought forth this favorable comment from a trade magazine: "It may be stated that the greatest hindrance to untrammeled progress in the development of better cracking, and still more to the general application of new developments to cracking technology throughout the country and the world has been lack of certainty as to just what principles were controlled by patents, and whether or not certain patents, considered basic in their principles, were valid or not."[32] This statement reveals that the main purpose of a pool may be to protect questionable patents. Preventing a clash of patents—a practice encouraged by the courts since it lessens litigation—may be used as a pretext for pooling or cross-licensing to avoid exposure of their invalidity and also to restrain trade.

To the extent that former competitors are brought into a patent pool, whatever the main reason may be, the chief source for challenging such patents is taken away. Members in accepting their validity may pre-empt the field through patent declaration rather than invention. Such a danger emphasizes again the need of a higher standard in the issue of patents and the need for alertness on the part of the Department of Justice in challenging patents in antitrust cases. Moreover, it indicates the soundness of pooling only the patents issued before a certain date, a practice illustrated in the automobile industry. This arrangement may settle pending litigation and yet fully preserve the incentive for judicial tests of validity and for inventive effort after that date.

[32] *National Petroleum News,* Dec. 8, 1937.

Instead of pooling rival patents that hinder the development or use of an invention for national defense, the government might purchase them, if valid, and make them available to all manufacturers. In 1918, the Senate Committee on Military Affairs proposed such a plan for aircraft patents. It stated, "New patents of meritorious character should be similarly acquired by the Government as they appear. Such a course would stimulate invention and promote the development of an art which is still in its cradle."[33] Seventeen years later, in 1935, General Mitchell said that he still agreed with this plan.[34]

Consolidations. In interpreting the Sherman Act the Supreme Court, though stating in the American Tobacco decision of 1911 and other cases that the acquisition of competitors is pertinent evidence, has never declared illegal any purchase of the patents, plants, or other assets of one company by another; but it has declared unlawful any restraint of trade between these same two companies. The effect of course is the concentration of patents and physical assets of each industry in fewer and larger companies. This in turn facilitates restraint of trade through leadership and collusion. Harmony in policy, in reference to patents, means no litigation to test infringement or validity of each other's patents. Such respect for private property permits corporations to preserve their patents and present a formidable front to would-be intruders. Corporations in effect pool their patents through comity rather than definite agreement and leave behind no tell-tale evidence. In 1950, Congress amended Section 7 of the Clayton Act to provide that no corporation shall acquire the assets of another if the effect "may be substantially to lessen competition or to tend to create a monopoly." Applying this amendment to patents as well as to physical assets is imperative to prevent the control of certain industries through concentration of patents.

There is need for vigorous competition in methods of communication, transportation, etc., as well as in the fabrication of chemicals and other products. Such rivalry does not exist in communication as in transportation. Would we permit the railways to control enough of the important patents on automotive and air transportation to retard their development? Also the Sherman Act prevents any regional

[33] Pooling Hearings, Part 1, 7.
[34] *Ibid.*, 820.

monopoly of railways, and the Panama Canal Act forbids the control of boat lines by competing railways. In the communication field, however, we permit American Telephone, through stock ownership, to dominate regional telephone companies and Western Electric to exercise considerable control over the telegraph and wireless by means of patents.

In the public-utilities industry, research has led incidentally to inventions in such private fields as motion-picture apparatus and medical equipment, but public utilities should be excluded from those private fields as manufacturers. It is unsound of course for the same owners to engage in public and private business—for example, the providing of telephone service and the manufacture and sale of telephone equipment. Certainly the railways are not permitted to engage in commercial manufacture in spite of fruitful research and prospective efficiency in such a venture. And gas and electric utilities generally neither manufacture appliances in these fields nor sell such merchandise made by others. Since the research by public utilities, whatever they may be, is paid for through utility rates, the resultant inventions should be available on a non-royalty basis.

License Agreements. Licenses that limit the "method of sales and the price" of former competitors were approved in 1926 in the General Electric decision[35] and then placed in doubt in the Line Material case of 1948.[36] As Justice Douglas pointed out in the second case, the Supreme Court is the author of this doctrine and should take the initiative in eliminating it.

In the Transparent-Wrap Machine decision of 1947, the Supreme Court held, five to four, that a licensor may legally require licensees to assign improvement patents to him, and then added that this restriction "could be employed with the purpose or effect of violating the antitrust laws."[37] The legal recognition of this right provides a means for indefinite control through patents because the users or manufacturers of an invention are the main source of improvements upon it. Until this weakness in our antitrust law is removed, the continued domination of some industries by one or a few companies will continue. Each licensee should have the right to keep the inventions, like

[35] 272 U.S. 476 (1926).
[36] 333 U.S. 287 (1948).
[37] 329 U.S. 637, 646 (1947).

other things, which he creates, and to develop a rival line of improvements. Awarding a patent to an inventor and encouraging competition are not inconsistent.

Cartels. Chapter Five emphasizes the part that patents play in forming and operating cartels. The first time the Supreme Court condemned a cartel was in 1947 in spite of the many years National Lead and other American companies had participated in a world-wide monopoly for the restraint of trade. There have been additional exposures of cartel activities since 1947 and others should follow if the government maintains its recent policy of vigorous enforcement of antitrust laws.

Since other nations generally do not oppose cartels, and in fact since some of them openly approve them, there is danger in any international organization for appraising cartels as good, bad, or indifferent. If and when this country ever makes the mistake of entering into such an arrangement, any effectiveness of the Department of Justice and the courts in enforcing our antitrust laws will be seriously impaired. This would be true even if this country reserved the right to enforce such legislation. There would arise the feeling and argument that since the cartel problem is being dealt with internationally—for example, by United Nations—this country, like other nations, should accept its interpretations and opinions. Antitrust action by the Department of Justice would put this country in the role of interfering and of refusing to co-operate. Any decision on the part of an international organization on the matter of our corporations' participation in cartels would be useless if it reached the same conclusion as our courts and would be worse than useless if it did not. In case of conflict, American judicial opinion, though prevailing, would be weakened.

It may be true that some cartels cannot be controlled by national processes. However, it is better to have effective action against even a few cartels than against none of them. Moreover, allowing suppression of patents by corporations and corporation violation of antitrust legislation to be a sufficient defense for the alleged infringement of patents should accomplish much in filling in the gap of inadequate jurisdiction against cartels in which corporations are involved.

The cartel, more than any other economic institution today, is closely tied to medieval mercantilism. The greater the importance of foreign trade in any country the stronger the support of the cartel. Even in a country where most sales are made to local residents there exists when selling abroad the willingness to deviate from the ideal of competition. The director of the export trade office of the Federal Trade Commission, stated in 1945 "that the effects of American combinations in restraint of export trade, if confined to such trade, are visited upon foreign competitors and upon the foreign buyer or consumer, rather than upon Americans."[38] Acceptance of such a doctrine, combined with insistence upon competition in the home market, requires legal differentiation between restraint of trade in foreign and in domestic commerce. The Webb-Pomerene Act, allegedly a defensive measure which enabled domestic manufacturers to combine in selling abroad in competition with export associations, and the like, of foreign nations, rests upon the naive assumption that restraint of trade may be separated into lawful and unlawful categories according to whether it is domestic or foreign. Fortunately the Supreme Court has held that participation in a cartel constitutes unfair competition and therefore is unlawful. Repealing the Webb-Pomerene law would strengthen our stand for competition at home and our opposition to cartels and kindred schemes abroad.

Our own government is partly to blame for perpetuating rather than destroying German domination in certain fields. It is true that some five thousand chemical patents seized during World War I were made available to our industries through the Chemical Foundation which purchased these patents for $270,000—a fact which helps to explain a six-fold increase between 1914 and 1939 in the value of our manufactured chemical products. On the other hand, according to Wendell Berge, in 1945, "After the last war the Germans were able to circumvent the policies of the Alien Property Custodian, with the consequence that within a decade after the war most of the important assets had been brought again within the sphere of either German influence or domination."[39] The custodian, for example, sold the Bayer and Bosch Magneto companies with 680 patents to United States interests which were later reacquired by the Germans. If these sales had been subject to reversion of the stock to the United States in case it ever fell into German hands—a policy adopted in the dis-

[38] *Hearings before Senate Committee on the Judiciary on S. 11,* 79 Cong., 1 sess., 73.
[39] EGRW Hearings, Part 4, 563.

posal of the stocks of Schering and American Bosch Magneto after World War II—German domination through patents would have been checked in two industries at least. Moreover, if other patents had been dedicated to the public or offered on a nonexclusive, royalty-free basis as they were during or after World War II, other rival industries might have developed in this country, and they might have remained free from cartel entanglements. The patents of American Bosch were made available, in 1942, through a consent decree and those of Schering and General Aniline and Film were opened to nonexclusive licensing by the Office of Alien Property after the war.

Control of Supplementary Products. Section 3 of the Clayton Act and the interpretations of this practice by the Supreme Court, beginning with the projecting-machine case in 1917,[40] leave little to be desired by those who would prevent encroachments upon competition in supplementary products by means of patents. However, as already explained in Chapter Six, these clear-cut victories in behalf of the public have been challenged by the new definition of contributory infringement in the Patent Act of 1952. Section 271 of this act, which contains the definition, should be repealed.

Forced Validity and Prolonged Monopoly. If and when the standard of patentability is administered strictly and consistently, invalid patents will become the exception rather than the rule, and therefore one purpose in avoiding a test of their validity will largely cease. Unfortunately, however, this proviso is not easily fulfilled. Only heroic and unpopular measures, as already described, can improve the quality of patents. There has been progress in exposing and even cancelling questionable patents by enforcing antitrust legislation. Licensees now have the recognized right to challenge the validity of patents if they enter into an agreement which would be illegal if unsupported by valid patents. In one case involving oil companies,[41] a district court condemned the companies' accepting the validity of each other's patents, and many other courts follow suit. Furthermore, the restoration of competition would bring back, in the words of Judge Simon H. Rifkind, "the chief spur to private resistance to the grant of monopolies."[42]

40 243 U.S. 502 (1917).
41 *United States* v. *Standard Oil Co. (Ind.)*, 33 F. 2d 617, 630 (N.D. Ill. 1929).
42 *United States* v. *National Lead Co.*, 63 F. Supp. 513, 532 (S.D.N.Y. 1945).

Prolonged monopoly may be checked partly by limiting the life of a patent to twenty years from the date of application—a remedy proposed by various groups but not yet even voted upon by Congress. This presupposes that the Patent Office would act expediently upon patent cases and that prospective patentees would proceed likewise either because of self-interest or because of compulsion. The Patent Office should make sure that a patent application reveals the way to practice the alleged invention; in antitrust suits the Department of Justice should challenge the validity of patents which are incompletely disclosed. In addition, any license agreement should cover only issued patents. To require the licensee to hand over his inventions perpetuates the licensor's dominance. In one instance at least, a court held that "because the grant-back provisions are inconsistent with the existence of 'effective competition,' . . . they should be decreed unenforceable."[43] Such a stand, it is hoped, will become general. Moreover, a stricter standard of patentability would eliminate many patents, perhaps most of them, and thus minimize or take away altogether the apparent right to continue a monopoly after the patents on important inventions have expired. The general availability of all patents arising from government-financed research would be another obvious check upon prolonged monopoly. This idea, recommended by the attorney general in 1946, has already been adopted for the inventions of government employees; in 1954, it was extended to research in atomic energy by business corporations if financed by the government.

Unfair Competition. The Federal Trade Commission has made little headway in checking unfair practices involving patent litigation because it is difficult to prove that the infringement of patents is intentional or that suits are brought in bad faith. A better means of checking abuses of this sort is for the Department of Justice to intercede in private litigation about patents if it appears that elimination of competition, rather than the adjudication of patent rights, is the objective. Many suits that have brought small competitors and users of the invention to terms would never have been threatened or started if the Antitrust Division could have challenged the purpose of the aggressor in the light of its past record. In 1944, for example,

[43] *United States* v. *Aluminum Company of America,* 91 F. Supp. 333, 410 (S.D.N.Y. 1950).

the Department of Justice intervened in the suit of Wisconsin Alumni Research Foundation against Rene Douglas for infringement of irradiation patents and converted the case into an antitrust suit; a consent judgment of 1946 cancelled the infringement suits, enjoined price fixing and other restrictions, and dedicated the patents to the public.[44] A more specific proposal in this connection is a law to forbid suits of infringement against the users of patented articles until infringement by the manufacturer of them has been definitely established.

Suppression of Patents. Possible remedies for willful suppression of patents include restoration of competition, establishing annual fees, and giving legal sanction to the recognition of suppression as a defense for infringement.

As previously explained, the domination of an industry through patents is often accompanied by the suppression of patents. In contrast, competition with an industry leads to rival lines of invention and the development and sale of them. Immediate profit is sacrificed if need be to keep abreast or ahead in quality of product. Familiar illustrations include refrigerators, gas ranges, and such gadgets as can openers. Moreover such industries would provide a competitive market for the patents of independent inventors; their interest would be in using rather than suppressing them. The emergence of alternate inventions from government research and their general availability should also hasten the return of competition in certain industries.

An annual tax on patents, as already proposed, would lead to the abandonment of many patents as the result of the refusal to pay the tax. Some of these patents might be on inventions regarded as inferior by the owners and yet attractive enough for improvement and development by beginners with an outlook free from vested interest. Ordinarily, however, industrial monopolies, which own most of the suppressed patents, would be willing and able to retain the patents on which their dominance depends.

The willful suppression of patents by corporations, depriving people of the opportunity to use and know the invention, should be made a sufficient defense for alleged infringement of them. In administering such a proposal, courts should define suppression as the acquir-

[44] CCH 1946–47 Trade Cases ¶ 57,433 (N.D. Ill. Civil No. 43–C–704; Complaint, 1944; Consent Decree, 1946).

ing of patents from employees or outsiders and after a reasonable period, say three or four years, failing to produce and sell the invention, in spite of sufficient finances, in an amount as great as the public would buy at a reasonable price. In determining what a reasonab' price would be attention should be given to what others would be willing to charge and to what would be a proper reward or return to the owner. As already indicated, the inventor-owner would be completely exempt. Such a defense for infringement would bring acquired and suppressed inventions into use, if they are worthwhile, either by the owner of the patents on them or by others. Self-interest and initiative in using such inventions would supplant self-interest and inertia in suppressing them. The result would be the promotion of inventions through development and use and perhaps the restoration of competition. Also, under this proposal, corporations might still take out patents on possible improvements and gadgets so that others could not patent them and demand royalties. Although the suppressed ones could be lawfully infringed, they would have served their original purpose of clearing a line of possible development without hindrance from outsiders. The argument that this proposal would bring inferior inventions into use ignores the likelihood that such inventions would not be infringed in the first place. A kindred contention, which applies especially to United States patents foreign-owned and suppressed, is the possible violation of the law of comparative costs—in other words, the likelihood of ill-adapted manufacturing in this country. However, such an industry would be temporary rather than permanent. Presumably enterprises would have the same self-interest and judgment in considering such an industry's possibilities as in making decisions about a business without any patents. Only a high tariff or an outright subsidy would result in its presence here.

Suppression as a defense for infringement should not apply, however, to the outside independent inventor during the time he keeps full legal command of his invention. The reason is that he, more than anybody else, seeks the development and use of his invention. Any suppression on his part usually results from inadequate finances rather than agreement or undue monopoly; permitting others to infringe his suppressed patent with impunity would impair his incentive to invent.

Enforcement. Within recent years, several judicial decisions and consent decrees involving the unlawful use of patents have required licenses with reasonable royalties. If this measure proves inadequate in restoring competition, the patents and trade-marks in question should be cancelled. This was done in the mimeograph decree of 1948 and in a district court decision of 1953 in reference to incandescent lamps. As Justice Douglas once stated in a dissenting opinion, "The impairment of property rights is no barrier to the fashioning of a decree which will grant effective relief. Divestiture or dissolution may be ordered in spite of hardship, inconvenience, or loss."[45]

Whatever the decision or decree, its effectiveness obviously would be greater if prosecution of offenders began soon after the violation of the law and continued without undue delay until final adjudication, and if vigilance were exercised in checking-up on the observance of the decision given. Carrying out such a program would require sustained initiative by the enforcement agencies and this in turn would call for enlarged staffs and adequate appropriations for them. There should be enough judges, of course, to keep the court calendars from being crowded. In the Hartford-Empire case alone the defendant spent an amount in excess of the annual appropriation for the Department of Justice. The Antitrust Division has not been able to take action in all justifiable complaints which have come to it.

Compulsory Licensing

Compulsory licensing, as the name indicates, is compelling a patent owner to let someone else use his invention. Such licensing has been employed since the early forties in judicial decisions and consent decrees involving the violation of antitrust laws. Also, as already noted, the government has the right to authorize any contractor to use any invention, the payment for which may be determined in the Court of Claims; the Atomic Energy Act provides for the exercise of similar power. Should compulsory licensing be extended to patent owners in general to prevent the suppression of worthwhile patents and to encourage competition? Such a proposal has its strongest support in reference to patents taken out in the United States by foreigners solely for the purpose of reserving this country as a market for their inventions manufactured abroad.

[45] *United States* v. *National Lead Co.*, 332 U.S. 319, 366 (1947).

Both England and Canada, along with many other countries, have compulsory licensing, but they have approved only a few patents to be so licensed. Opponents of such a measure for the United States emphasize this fact and argue therefore that it is not needed. However, the importance of the action cannot be measured by the mere number to whom compulsory licenses are granted. Evidence indicates that the mere existence of compulsory licensing induces patentees to adopt a reasonable attitude in dealing with requests for voluntary licenses. An agreement, for example, between Siemens and Halske of Germany and Beryllium Corporation of the United States which barred American sales in England, was changed because of a threat to apply the compulsory licensing law to the German concern.[46] Incidentally, another factor conducive to voluntary licenses in England is the legal right of the patentee to impose price and other restrictions upon licensees which enable the licensor to retain control.

The greatest drawback to compulsory licensing lies in its administration. It would be difficult to appraise the relative merit or importance of kindred patents in the same industry which were owned by various individuals and companies; yet this would be necessary in determining the royalties for each owner. Furthermore, appraisals of patents would be modified as some patents expire and others are issued. Also, a manufacturer in another industry might want a license under one or a few of these patents rather than all of them. Improvement patents would have to be differentiated and the royalty determined for each separately so that any licensee could take any one or all of them. Also, the proper administration of compulsory licenses would require appropriate action to test the validity of the patents in question.

Two previous proposals—letting the suppression of patents be a sufficient defense for infringement and preventing the consolidation of patents in one industry—would accomplish practically all that compulsory licensing is supposed to bring about and do it more easily and completely. For this reason compulsory licensing should be used only as a second alternative and, therefore, only in case these two previous proposals cannot be put into effect.

[46] TNEC Report, 9.

Government-financed Research and Invention

Government-financed research may become still more important in the future. Its outstanding objectives at present are national defense and public health. Another goal should be the restoration of competition in those industries in which dominance rests upon undue concentration of patents. Government patents may do to such industries what the parcel post of 1910 did to the express business.

Can the Atomic Energy Commission, Research and Development Board, and the National Science Foundation measure up to the Bureau of Standards in quality of work? Will the research of the government, like that of private corporations, consist of intensive improvements and additional applications rather than revolutionary discoveries? Apparently a fruitful source of basic invention will still be university professors like Henry and Babcock—trained in science and free from the need of commercial application. The government should draw upon this source to the utmost. Of course, projects farmed out to private corporations will remain significant.

According to a presidential order, in 1950, the government shall have the entire right to the inventions of its employees, excepting of course the ones which they make on their own time and at their own expense. This is a highly commendable change in policy. The government and its citizens are entitled to the services they pay for. More important, the ownership of patent rights is not a necessary or proper incentive to government scientists and technicians. In government service as in private industry such rights detract from team work, encourage secretiveness, and unduly emphasize patentability. Furthermore, meritorious service of all government employees should be rewarded alike through promotion and special recognition.

This patent policy should apply also to the research of professors and other individuals affiliated with universities and other nonprofit institutions which are financed by the government. Apparently there is no causal relationship between their great inventions and prospective money-making from patents. In fact, some of them, though having the right to take out and exploit patents, have refused to do so. Incidentally, several universities now carry out this policy by taking the patents on inventions arising from the research which they finance; they then in turn dedicate them to the public.

Most important of all is the need of extending this policy of the government's acquiring the inventions it pays for to government-financed research carried out by private corporations. Within recent years the government, particularly the army and navy, have paid millions to corporations for research and yet have let them have the patents on the attendant inventions. The government retained only shop rights—in effect, a royalty-free license for its own use only. The situation should be reversed—patents to the government and a shop right to the corporations. This right plus acquired know-how, in addition to payment for the research itself, would be sufficient inducement to them.

Under the present practice the companies acquire control, through patents, of the commercial applications of the research done for the government. Though dominant already, they obtain additional patents to strengthen and perpetuate their positions. In this manner, the Department of Defense subsidizes the building up of monopolies which the Department of Justice seeks to destroy by enforcing anti-trust laws, and it partially nullifies the government programs for encouraging small enterprises by loans and other means. Moreover, letting a company keep the patents does not encourage its inventors or technicians, for they receive salaries rather than patents for their research. These patents and the know-how that goes with them, financed with public funds, should go to the government and then be made available to citizens of this country. An excellent beginning in this direction is the law of 1954 which requires that the results from government-financed atomic research must become public property.

In administering patents which result from research financed by the government or which come from the alien property custodian, the underlying policy should be to make them available to nationals by dedication or by royalty-free, nonexclusive licensing. Of course, important inventions for national defense should still be retained in secrecy until proper authorities sanction their publication and perhaps dedication. Furthermore, when the government acquires title to the patents of its employees, it should also secure ownership of the foreign rights as it is now with few exceptions required to do.[47] More important, the government should continue to have the foreign as well as the United States patents arising from the research of uni-

[47] Executive Order 9865.

versities and private corporations which government financing has made possible. Such a policy now helps domestic enterprises to get into foreign markets and offsets any practice of foreign nations in taking out patents in this country. Such a policy may result in inventions and know-how sufficiently important to encourage other nations with government-owned patents to enter into agreement for the international exchange of government-owned patents—an arrangement which the United States has already proposed to other governments.

The government, in its research program, should not overlook the independent inventor. The unexpected and the upsetting will come from him in the future as in the past. Also, he will author more than his share of worthless inventions. Nevertheless, the government should encourage him to come to the appropriate agency and there present his invention and its possibilities. During the recent war the National Inventors Council screened many inventions for the armed services and some of the inventions thus screened were accepted for military use. Inventors should be encouraged to use this service to a greater extent in time of peace. Also, the government should let inventors know about organizations whose business is to find a market for promising inventions in return for a part of the royalties. A higher standard in granting patents and the restoration of competition in certain industries would be helpful in selling patents whether to the government or to private enterprise.

The patent law now provides, as stated in Chapter One, that if any application for a patent on an invention relates to national security, the government may keep it secret and use it upon the condition that the government make full settlement, upon request of applicant, for damages rising from such action. However, if any other invention is used or manufactured by the government or by its contractors, the patentee's means for securing compensation, in case of refusal of payment by the government agency, is limited to action against the government in the courts. In this instance, as in the previous situation, the government should offer the patentee what it deems a proper settlement and then give him the choice of accepting it or appealing to a board or commission. Any award within reason would be far better than the mere right to litigate.

Conclusion

Our patent system has undergone significant changes in the course of its history. In large measure, the hired inventor has supplanted the independent inventor and, therefore, salary has taken the place of the patent as remuneration. Another change is the growing extent to which patents are invalid and to which schemes are used to avoid their exposure. A third change is the rising proportion of patented inventions never put into practice. A fourth is the increasingly prohibitive cost of litigation. Last but not least, is the increased exploitation of patents to defeat antitrust laws.

Since our patent system as a means for encouraging and obtaining invention is now effective primarily in the case of the independent inventor, proposed remedies should be designed for him. This calls, first, for a distinction between the independent inventor's patents and those of the hired inventor. Second, there should be equal vigilance in rejecting and in accepting applications for patents; acceptance should be made in accordance with a high standard of patentability. Third, to the extent that invalid and worthless patents still emerge from the Patent Office, they should be eliminated by a challenging of their validity and by imposing annual fees. Fourth, to lessen conflict and delay and expense, the life of a patent should be limited to twenty years from the date of application and a board or commission provided for determining compensation to inventors in return for the use of their inventions by the government and its contractors. Fifth, proof of suppression of patents should be a sufficient defense for alleged infringement. Another imperative is prosecution under antitrust laws that wins decisions and then enforces the decisions. These six proposals would go a long way in removing the evils of our patent system. Moreover, they would not affect the extent to which inventions are kept secret. Hiding them in the future, as in the past, will depend upon their nature—upon whether they are a process or apparatus which can be used by the owner and still withheld from others—rather than a patent system. It will always be easier to be vigilant in one's own shop or factory than in the entire nation.

The prospect of the adoption of any of these measures by Congress is very remote. So far at least, our lawmakers have not responded to numerous hearings and reports and decisions by committees and commissions and courts that reveal the evils of our patent system and suggest certain changes. The lawmakers will not put into effect even the simple proposal, generally recommended, that a patent shall not extend beyond twenty years from the date of application. Apparently, the only hope for improvement lies in a judicial interpretation which will restore the original purposes of our patent and antitrust laws—to promote the useful arts by securing for limited times to *inventors* the exclusive rights to their respective discoveries and to prevent industrial monopoly and restraint of trade. Progress in this direction has been made by the courts in recent years. Much more is needed.

Abbreviations

Used in References to Hearings and Reports

Bone Hearings	*Patents, Hearings before Senate Committee on Patents on Senate Resolutions 2303 and 2491, Parts 1–9,* 77 Cong., 2 sess.
CCH, F. A. L. Case—	*The Federal Antitrust Laws with Summary of Cases, 1890–1951.* Washington, Commerce Clearing House, 1952.
Comp. L. Hearings	*Compulsory Licensing of Patents, Hearings before House Committee on Patents on House Resolutions 9259, 9815, and 1666,* 75 Cong., 3 sess.
C S B Report	*United States versus Economic Concentration and Monopoly,* 79 Cong., 2 sess., *House Resolution No. 64.*
EGRW Hearings	*Elimination of German Resources for War, Hearings before Senate Committee on Military Affairs on Senate Resolutions 107 and 146, Parts 1, 2, 3, 4, 10,* 79 Cong., 1 and 2 sess.
F C C Report	*Investigation of the Telephone Industry in the United States,* 76 Cong., 1 sess., *House Report No. 340.*
M P Hearings	*Study of Monopoly Power, Hearings before House Committee on House Resolutions 137, 156, 6679, and 7905, Serial 14, Parts 1, 2–A, 2–B, 3, 4–A, 4–B, 5, 6–A, 6–B,* 81 Cong., 1 and 2 sess.
Nolan Hearings of 1919	*United States Patent Office, Hearings before House Committee on House Resolutions 5011, 5012, and 7010,* 66 Cong., 1 sess.
N R C Report	*Technological Trends and National Policy,* 75 Cong., 1 sess., *House Report No. 360.*

Oldfield Hearings of 1912	*Revision and Codification of the Patent Statutes, Hearings before House Committee on Patents on House Resolution 23417, Parts 1–27,* 62 Cong., 2 sess.
Oldfield Hearings of 1914	*Revision and Codification of the Patent Statutes, Hearings before House Committee on Patents on House Resolution 15989, Parts 1–10,* 63 Cong., 2 sess.
Oldfield Report of 1912	62 Cong., 2 sess., *House Report No. 1161.*
Oldfield Report of 1914	63 Cong., 2 sess., *House Report No. 1082.*
Pooling Hearings	*Pooling of Patents, Hearings before House Committee on Patents on House Resolution 4523, Parts I–IV,* 74 Cong.
S T M Hearings	*Scientific and Technical Mobilization, Hearings before Senate Committee on Military Affairs on Senate Resolutions 107 and 702, Part IV (Patents),* 78 Cong., 1 sess.
TNEC Hearings	*Investigation of Concentration of Economic Power, Hearings before Temporary National Economic Committee on Public Resolution 113, Parts 2 and 3 (Patents) and Part 25 (Cartels),* 75 Cong.
TNEC Report	*Investigation of Concentration of Economic Power,* 76 Cong., 1 sess., *House Report No. 95.*

Bibliography

Court Decisions and Consent Decrees

Allen Calculators v. *National Cash Register Co.*, 322 U.S. 137 (1944).

B. B. Chemical Co. v. *Ellis*, 314 U.S. 495 (1942).

Bauer v. *O'Donnell*, 229 U.S. 1 (1913).

Bement v. *National Harrow Co.*, 186 U.S. 70 (1902).

Bereslavsky v. *Esso Standard Oil Co.*, 175 F. 2d 148 (4th Cir. 1949).

Bereslavsky v. *Standard Oil Co.* (*N. J.*), 82 F. Supp. 939 (D. Md. 1949).

Besser Mfg. Co. v. *United States*, 343 U.S. 444 (1952).

Bloomer v. *McQuewan*, 14 How. 539 (1852).

Bloomer v. *Millinger*, I Wall. 340 (1863).

Blount Mfg. Co. v. *Yale and Towne Mfg. Co.*, 166 Fed. 555 (C.C. Mass. 1909).

Bobbs-Merrill Co. v. *Straus*, 210 U.S. 339 (1908).

Boston Store v. *American Graphophone Co.*, 246 U. S. 8 (1918).

Carbice Corp. v. *American Patents Development Corp.*, 283 U.S. 420 (1931).

Carbice Corp. v. *American Patents Development Corp.*, 283 U.S. 27 (1931).

Chaffee v. *Boston Belting Co.*, 22 How. 217 (1859).

Coke, 3 Institutes 181.

Columbia Motor Car Co. v. *Duerr and Co.*, 184 Fed. 893 (2nd Cir. 1911).

Continental Paper Bag Co. v. *Eastern Paper Bag Co.*, 150 Fed. 741 (1st Cir. 1906); 210 U.S. 405 (1908).

Cortelyou v. *Johnson*, 207 U.S. 196 (1907).

Cotton-Tie Co. v. *Simmons*, 106 U.S. 89 (1882).

Cuno Engineering Corp. v. *Automatic Devices Corp.*, 314 U.S. 84 (1941).

Darcy v. *Allin*, 11 Coke 84b (K. B. 1602).

Densmore v. Scofield, 102 U.S. 375 (1880).

Dr. Miles Medical Co. v. *Park & Sons*, 220 U.S. 373 (1911).

Ethyl Gasoline Corp. v. *United States*, 309 U.S. 436 (1940).

Ewart Mfg. Co. v. *Baldwin Cycle-chain Co.*, 91 Fed. 262 (C.C.D. Mass. 1898).

Federal Trade Commission v. *Eastman Kodak Co.*, 274 U.S. 619 (1927).

Ferguson v. *Ford Motor Co.* (S.D.N.Y. Civil No. 44–482; Complaint, 1948; Consent Judgment, 1952); also 77 F. Supp. 425 (S.D.N.Y. 1948).

Flynn and Emrich Co. v. *Federal Trade Commission,* 52 F. 2d 836 (4th Cir. 1931).

General Electric Co. v. *Willey's Carbide Tool Co.,* 33 F. Supp. 969 (E.D. Mich. 1940).

General Talking Pictures Corp. v. *Western Electric Co.,* 305 U.S. 124 (1938).

Gibbons v. *Ogden,* 9 Wheaton 1 (1824).

Hartford-Empire Co. v. *United States,* 323 U.S. 386 (1945); 324 U.S. 570 (1945).

Hazel-Atlas Glass Co. v. *Hartford-Empire Co.,* 322 U.S. 238 (1944).

Heaton-Peninsular Button Fastener Co. v. *Eureka Specialty Co.,* 77 Fed. 288 (6th Cir. 1896).

Henry v. *A. B. Dick Co.,* 224 U.S. 1 (1912).

Heuser v. *Federal Trade Commission,* 4 F. 2d 632 (7th Cir. 1925).

Heyer v. *Duplicator Mfg. Co.,* 263 U.S. 100 (1923).

Hoe v. *Knap,* 27 Fed. 204 (C.C.N.D. Ill. 1886).

Hotchkiss v. *Greenwood,* 52 U.S. 248 (1850).

Hughes Tool Co. v. *Williams,* 82 F. Supp. 408 (1949).

Indiana Mfg. Co. v. *J. I. Case Threshing Machine Co.,* 148 Fed. 21 (C.C.E.D. Wis. 1906); 154 Fed. 365 (7th Cir. 1907); 207 U.S. 603 (1907).

International Business Machines Corp. v. *United States,* 298 U.S. 131 (1936).

International Salt Co. v. *United States,* 332 U.S. 392 (1947).

Interstate Circuit, Inc. v. *United States,* 306 U.S. 208 (1939).

Katzinger Co. v. *Chicago Metallic Mfg. Co.,* 329 U.S. 394 (1947).

Kendall v. *Windsor,* 21 How. 322 (1858).

Keystone Steel and Wire Co. v. *Commissioner of Internal Revenue,* 62 Fed. 2d 458 (7th Cir. 1932).

Leeds and Catlin v. *Victor Talking Machine Co.,* 213 U.S. 325 (1909).

Leitch Mfg. Co. v. *Barber Co.,* 302 U.S. 458 (1938).

Lynch v. *Magnavox Co.,* 94 F. 2d 883 (9th Cir. 1938).

MacGregor v. *Westinghouse Electric and Mfg. Co.,* 329 U.S. 402 (1947).

Mercoid Corp. v. *Mid-Continent Investment Co.,* 320 U.S. 661 (1944).

Mercoid Corp. v. *Minneapolis-Honeywell Regulator Co.,* 320 U. S. 680 (1944).

Mitchell v. *Hawley,* 16 Wall. 544 (1872).

Morgan Envelope Co. v. *Albany Paper Co.,* 152 U.S. 425 (1894).

Morton Salt Co. v. *Suppiger Co.,* 314 U.S. 488 (1942).

Motion Picture Patents Co. v. *Universal Film Mfg. Co.,* 235 Fed. 398 (2nd Cir. 1916); 243 U.S. 502 (1917).

National Harrow Co. v. *Bement,* 47 N.Y.S. 462 (1897).

National Harrow Co. v. *Hench,* 83 Fed. 36 (3rd Cir. 1897).

National Phonograph Co. v. *Schlegel,* 128 Fed. 733 (8th Cir. 1904).

Picard v. *United Aircraft Corp.,* 128 F. 2d 632 (2nd Cir. 1942).

Pope Mfg. Co. v. *Gormully,* 144 U.S. 224 (1892).

Potts v. *Coe,* 140 F. 2d 470 (D.C. Cir. 1944).

Precision Instrument Mfg. Co. v. *Automotive Maintenance Mach. Co.,* 324 U.S. 806 (1945).

Richmond Screw Anchor Co. v. *United States,* 275 U.S. 331 (1928).

Rubber Tire Wheel Co. v. *Milwaukee Rubber Works Co.,* 142 Fed. 531 (C.C.E.D. Wis. 1906); 154 Fed. 358 (7th Cir. 1907); 210 U.S. 439 (1908).

Sinclair and Carroll Co. v. *Interchemical Corp.,* 325 U.S. 327 (1945).

Singer Mfg. Co. v. *June Mfg. Co.,* 163 U.S. 169 (1896).

Solo Electric Co. v. *Jefferson Electric Co.,* 317 U.S. 173 (1942).

Special Equipment Co. v. *Coe,* 144 F. 2d 497 (D.C. Cir. 1944); and 324 U.S. 370 (1945).

Standard Oil Co. (Ind.) v. *United States,* 283 U.S. 163 (1931).
(2nd Cir. 1946); 161 F. 2d 565 (2nd Cir. 1947).

Standard Oil Co. (Ind.) v. *United States,* 283 U.S. 163 (1931).

Standard Sanitary Mfg. Co. v. *United States,* 226 U.S. 20 (1912).

Stewart-Warner Corp. v. *Staley,* 42 F. Supp. 140 (W.D. Pa. 1941).

Stokes and Smith Co. v. *Transparent-Wrap Machine Corp.,* 156 F. 2d 198

Straus v. *American Publishers' Association,* 231 U.S. 222 (1913).

Straus v. *Victor Talking Machine Co.,* 243 U.S. 490 (April 9, 1917).

Technical Tape Corp. v. *Minnesota Mining and Mfg. Co.,* 200 F. 2d 876 (2nd Cir. 1952).

Times-Picayune Publishing Co. v. *United States,* 345 U.S. 594 (1953).

Timpkin Roller Bearing Co. v. *United States,* 341 U.S. 593 (1951).

Transparent-Wrap Machine Corp. v. *Stokes and Smith Co.,* 329 U.S. 637 (1947).

United Shoe Machinery Corp. v. *United States,* 258 U.S. 451 (1922).

United States v. *Aerofin Corp.,* CCH Trade Regulation Reports, Supp. 1941–43 ¶ 52,913 (S.D.N.Y. Civil No. 20–458; Complaint and Consent Decree, Mar. 5, 1943).

United States v. *Alba Pharmaceutical Co.,* CCH Trade Regulation Reports, Supp. 1941–43 ¶ 52,650 (S.D.N.Y. Civil No. 15–363; Complaint and Consent Decree, Sept. 5, 1941).

United States v. *Allegheny Ludlum Steel Corp.,* CCH 1948–49 Trade Cases

¶ 62,330 (D.N.J. Civil No. 45–83; Complaint, 1945; Consent Judgment, 1948).

United States v. *Aluminum Co. of America,* CCH Trade Regulation Reports, Supp. 1941–43 ¶ 52,776 (S.D.N.Y. Civil No. 18–31; Complaint and Consent Decree, April 15, 1942).

United States v. *Aluminum Co. of America,* 44 F. Supp. 97 (S.D.N.Y. 1941); 148 F. 2d 416 (2nd Cir. 1945).

United States v. *Aluminum Co. of America,* 91 F. Supp. 333 (S.D.N.Y. 1950).

United States v. *American Air Filter Co.,* CCH 1946–47 Trade Cases ¶ 57,492 (W.D. Ky. Civil No. 574; Complaint, 1943; Consent Judgment, 1946).

United States v. *American Bosch Corp.,* CCH 1948–49 Trade Cases ¶ 62,284 (S.D.N.Y. Civil No. 20–164; Complaint and Consent Decree, Dec. 29, 1942).

United States v. *American Can Co.,* 230 Fed. 859 (D. Md. 1916).

United States v. *American Can Co.,* 87 F. Supp. 18 (N. D. Calif. 1949).

United States v. *American Can Co.,* CCH 1950–51 Trade Cases ¶ 62,679 (N.D. Calif. Civil No. 26345–H; Complaint, 1946; Final Decree, 1950); *ibid.* (N.D. Calif. Cr. No. 34089; Complaint, 1954).

United States v. *American Lecithin Co.,* CCH 1946–47 Trade Cases ¶ 57,542 (N.D. Ohio Civil No. 24115; Complaint, 1946; Consent Judgment, 1947).

United States v. *American Locomotive Co.,* CCH 1945–47 Trade Cases ¶ 57,621 (N.D. Ind. Civil No. 545; Complaint, 1942; Consent Judgment, 1947).

United States v. *American Optical Co.,* CCH 1948–49 Trade Cases ¶ 62,308 (S.D.N.Y. Civil No. 10–391; Complaint, 1940; Consent Judgment, 1948).

United States v. *Aqua Systems,* CCH Trade Regulation Reports, Supp. 1941–43 ¶ 52,880 (S.D.N.Y. Civil No. 19–516; Complaint, 1942; Consent Decrees, 1942).

United States v. *Auditorium Conditioning Corp.,* CCH 1944–45 Trade Cases ¶ 57,428; (S.D.N.Y. Civil No. 22–200; Complaint, 1943; Final Judgment, 1945).

United States v. *Austenal Laboratories,* CCH 1950–51 Trade Cases ¶ 62,880 (S.D.N.Y. Civil No. 50–496; Complaint, 1949; Consent Decree, 1951).

United States v. *Automatic Sprinkler Co.,* CCH 1948–49 Trade Cases ¶ 62,230 (N.D. Ill. Civil No. 46–C–1289; Complaint, 1946; Consent Judgment, 1948).

United States v. *Bausch and Lomb Optical Co.,* CCH Trade Regulation Reports, Supp. 8th ed. Vol. 3 ¶ 25,487 (S.D.N.Y. Civil No. 9–404; Complaint and Consent Decree, July 9, 1940).

United Sates v. *Bausch and Lomb Optical Co.,* 321 U.S. 707 (1944).

United States v. *Bearing Distributors Co.,* CCH 1952–53 Trade Cases ¶ 67,595 (W.D. Mo. Civil No. 6895; Complaint, 1951; Consent Decree, 1953).

United States v. *Bendix Aviation Corp.,* CCH 1946–47 Trade Cases ¶ 57,444 (D.C.N.J. Civil No. 2531; Complaint, 1942; Consent Decree, 1946).

United States v. *Bendix Aviation Corp., Westinghouse Air Brake Co., et al.,* CCH 1948–49 Trade Cases ¶ 62,349 (S.D.N.Y. Civil No. 44–284; Complaint, 1947; Consent Judgment *re* Westinghouse, 1948).

United States v. *Besser Mfg. Co.,* 96 F. Supp. 304 (E.D. Mich. 1951).

United States v. *Cincinnati Milling Machine Co.,* CCH 1953–54 Trade Cases ¶ 67,733 (E.D. Mich. Civil No. 13401; Complaint and Consent Judgment, April 19, 1954).

United States v. *Consolidated Car-Heating Co.,* CCH 1950–51 Trade Cases ¶ 62,656 (S.D.N.Y. Civil No. 34–312; Complaint, 1945; Summary Judgment, 1950).

United States v. *Corning Glass Works* (S.D.N.Y. Cr. No. 108–164; Indictment, 1940; Fines, 1941).

United States v. *Davis Co.,* CCH 1952 Trade Cases ¶ 67,403 (S.D.N.Y. Civil No. 54–357; Complaint, 1949; Consent Judgment, 1952).

United States v. *Descher,* 1 D & J. 645 (1917).

United States v. *Diamond Match Co.,* CCH 1948–49 Trade Cases ¶ 57,456 (S.D.N.Y. Civil No. 25–397; Complaint, 1944; Consent Decree, 1946).

United States v. *A. B. Dick Co.,* CCH 1948–49 Trade Cases ¶ 62,233 (N.D. Ohio Civil No. 24188; Complaint, 1946; Consent Judgment, 1948).

United States v. *Dubilier Condenser Corp.,* 289 U.S. 178 (1933).

United States v. *E. I. Du Pont de Nemours & Co.,* 118 F. Supp. 41 (D. Del. 1953).

United States v. *Eastman Kodak Co.,* 226 Fed. 62 (W.D.N.Y. 1915); 230 Fed. 522 (W.D.N.Y. 1916); 255 U.S. 578 (1921).

United States v. *Eastman Kodak Co.,* Equity A–51, Final Decree (W.D. N.Y. (1921).

United States v. *Electric Storage Battery Co., and Willard Storage Battery Co.,* CCH 1946–47 Trade Cases ¶ 57,645 (S.D.N.Y. Civil No. 31–225; Complaint, 1945; Consent Judgment, 1947).

United States v. *Gamewell Co.,* CCH 1948–49 Trade Cases ¶ 62,236 (D. Mass. Civil No. 6150; Complaint, 1946; Consent Judgment, 1948).

United States v. *Gamewell* Co., 95 F. Supp. 9 (D. Mass. 1951).

United States v. *General Cable Corp.,* CCH 1948–49 Trade Cases ¶ 62,300 (S.D.N.Y. Civil No. 40–76; Complaint, 1947; Consent Judgment, 1948).

United States v. *General Electric Co.,* 1 D & J. 267 (1911).

United States v. *General Electric Co.,* 272 U.S. 476 (1926).

United States v. *General Electric Co., Westinghouse Electric and Mfg. Co., et al.,* CCH Trade Regulation Reports, Supp. 1941–43, ¶ 52,777 (D.N.J. Civil No. 1364; Complaint, 1941; Consent Decree *re* Westinghouse, 1942).

United States v. *General Electric Co., Corning Glass Works, et al.,* CCH 1946–47 Trade Cases ¶ 57,448 (D.N.J. Civil No. 2590; Complaint, 1942; Consent Decree *re* Corning, 1946).

United States v. *General Electric Co.,* 80 F. Supp. 989 (S.D.N.Y. 1948).

United States v. *General Electric Co.,* 82 F. Supp. 753 (D.N.J. 1949).

United States v. *General Electric Co.,* 115 F. Supp. 835 (D.N.J. 1953).

United States v. *General Electric Co.,* CCH 1952–53 Trade Cases ¶ 67,585 (D.C.N.J. Civil No. 45–75–C; Complaint, 1945; Final Consent Judgment, 1953).

United States v. *General Electric Co.,* CCH 1953–54 Trade Cases ¶ 67,794–67,796 (D.N.J. Civil No. 2590; Complaint, 1942; Consent Decree *re* General Electric *et al.,* 1954).

United States v. *General Instrument Corp.,* 87 F. Supp. 157 (D.N.J. 1949); 115 F. Supp. 582 (D.N.J. 1953).

United States v. *Hartford-Empire Co.,* 46 F. Supp. 541 (N.D. Ohio 1942).

United States v. *Imperial Chemical Industries, Ltd.,* 100 F. Supp. 504 (S.D.N.Y. 1951).

United States v. *Krentler-Arnold Hinge Last Co.,* 1 D & J. 407 (1913).

United States v. *Libby-Owens-Ford Glass Co., et al.,* CCH 1948–49 Trade Cases ¶ 62,323 (N.D. Ohio Civil No. 5239; Complaint, 1945; Final Decree *re* Corning, 1946; Consent Judgment *re* others, 1948).

United States v. *Line Material Co.,* 333 U.S. 287 (1948).

United States v. *The Liquid Carbonic Corp.,* CCH Trade Cases ¶ 67,248 (E.D.N.Y. Civil No. 9179; Complaint, 1948; Final Judgment, 1952).

United States v. *Mager and Gougelman, Inc.,* CCH 1951–52 Trade Cases ¶ 67,233 (N.D. Ill. Civil No. 49–C–1028; Complaint, 1949; Consent Judgment, 1952).

United States v. *Masonite Corp.,* 316 U.S. 265 (1942).

United States v. *Merck & Co.,* CCH 1944–45 Trade Cases ¶ 57,416 (D.C.N.J. Civil No. 3159; Complaint, 1943; Consent Decree, 1945).

United States v. *Minnesota Mining and Mfg. Co.,* 92 F. Supp. 947 (D. Mass. 1950); 96 F. Supp. 356 (D. Mass. 1951).

United States v. *Morton Gregory Corp.*, CCH 1950–51 Trade Cases ¶ 62,750 (N.D. Ohio Civil No. 6279; Complaint, 1949; Consent Judgment, 1951).

United States v. *Motion Picture Patents Co.*, 225 Fed. 800 (E.D. Pa. 1915).

United States v. *National Cash Register Co.*, CCH Trade Regulation Reports, Ct. Dec. Supp. III, ¶ 3,130 (C.C.S.D. Ohio Eq. No. 6802; Petition, 1911; Consent Decree, 1916).

United States v. *National Cash Register Co.*, (S.D. Ohio Cr. No. 7092; Petition and Consent Judgment, Jan. 8, 1947).

United States v. *National Lead Co.*, 63 F. Supp. 513 (S.D.N.Y. 1945); *ibid.*, 332 U.S. 319 (1947).

United States v. *New Departure Mfg. Co.*, 204 Fed. 107 (W.D.N.Y. 1913).

United States v. *New Wrinkle, Inc.*, 342 U.S. 371 (1952).

United States v. *Owens-Corning Fiberglass Corp.*, CCH 1948–49 Trade Cases ¶ 62,442 (N.D. Ohio Civil No. 5778; Complaint, 1947; Consent Judgment, 1949).

United States v. *Owens-Illinois Glass Co.*, CCH 1946–47 Trade Cases ¶ 57,498 (N.D. Calif. Civil No. 25861–G; Complaint, 1946; Consent Judgment, 1946).

United States v. *Parke, Davis and Co. and Eli Lilly and Co.*, CCH 1950–51 Trade Cases ¶ 62,914 (E.D. Mich. Civil No. 8940; Complaint, 1950; Consent Judgment, 1951).

United States v. *Parker Rust Proof Co.*, 61 F. Supp. 805 (E.D. Mich. 1945).

United States v. *Patent Button Co.*, CCH 1946–47 Trade Cases ¶ 57,579 (D.C. Conn. Civil No. 1854; Complaint, 1946; Consent Judgment, 1947).

United States v. *Permutit Co.*, CCH 1950–51 Trade Cases ¶ 62,888 (S.D.N.Y. Civil No. 32–394; Complaint, 1945; Consent Judgment, 1951).

United States v. *Phillips Screw Co.*, CCH 1948–49 Trade Cases ¶ 62,394 (N.D. Ill. Civil No. 47–C–147; Complaint, 1947; Consent Judgment, 1949).

United States v. *Radio Corp. of America*, CCH 1932–37 Trade Regulation Reports, Ct. Dec. Supp. ¶ 7,025 (D. Del. Eq. No. 793; Petition, 1930; Consent Decree, 1932).

United States v. *Radio Corp. of America*, 46 F. Supp. 654 (D. Del. 1942).

United States v. *Rail Joint Co.*, CCH 1946–47 Trade Cases ¶ 57,469 (N.D. Ill. Civil No. 43–C–1295; Complaint, 1943; Consent Judgments, 1944 and 1946).

United States v. *Rohm and Haas Co.*, CCH 1948–49 Trade Cases ¶ 62,334 (E.D. Pa. Civil No. 9068; Complaint and Consent Judgment, Nov. 18, 1948).

United States v. *Sand Spun Patents Corp.*, CCH 1948–49 Trade Cases ¶

62,462 (D.N.J. Civil No. 125–49; Complaint, 1949; Consent Judgment, 1949).

United States v. *Schering Corp.,* CCH Trade Regulation Reports, Supp. 1941–43, ¶ 52,729 (D.C.N.J. Civil No. 1919; Complaint and Consent Decree, Dec. 17, 1941).

United States v. *Dietrich A. Schmitz,* CCH 1944–45 Trade Cases ¶ 57,305 (S.D.N.Y. Cr. No. 111–137; Indictment, 1941; Imposition of Fines, 1950).

United States v. *Scophony Corp.,* 69 F. Supp. 666 (S.D.N.Y. 1946); 333 U.S. 795 (1948).

United States v. *Scophony Corp.,* CCH 1948–49 Trade Cases ¶ 62,238, ¶ 62,356 and ¶ 62,463 (S.D.N.Y. Civil No. 34–184; Complaint, 1945; Consent Decrees, 1948 and 1949).

United States v. *Scott and Williams, Inc.,* CCH 1953–54 Trade Cases ¶ 67,748 (S.D.N.Y. Civil No. 54–358; Complaint, 1949; Consent Decree, 1954).

United States v. *Scovill Mfg. Co.,* CCH 1948–49 Trade Cases ¶ 62,223 (D.C. Conn. Civil No. 1853; Complaint, 1946; Consent Judgment, 1948).

United States v. *Servel, Inc.,* CCH 1953–54 Trade Cases ¶ 67,665 (E.D. Pa. Civil No. 11036; Complaint, 1950; Consent Judgment, 1954).

United States v. *Standard Oil Co.* (*Ind.*), 33 F. 2d 617 (N.D. Ill. 1929).

United States v. *Standard Oil Co.* (*N.J.*), CCH Trade Regulation Reports, Supp. 1941–43 ¶ 52,768 and ¶ 52,927 (D.C.N.J. Civil No. 2091; Complaint and Consent Decree, March 25, 1942; Supp. Judgment, 1943).

United States v. *Standard Register Co.,* CCH 1950–51 Trade Cases ¶ 62,533 (D.C. of D.C. Civil No. 36040; Complaint, 1946; Consent Judgment, 1949).

United States v. *Standard Sanitary Mfg. Co.,* 191 Fed. 172 (C.C. Md. 1911).

United States v. *Switzer Brothers, Inc.,* CCH 1952–53 Trade Cases ¶ 67,605 (N.D. Calif. Civil No. 29860; Complaint, 1950; Consent Judgments, 1953).

United States v. *Synthetic Nitrogen Products Corp.,* CCH Trade Regulation Reports, Supp. 1941–43 ¶ 52,700 (S.D.N.Y. Civil No. 15–365; Complaint and Consent Decree, Sept. 5, 1941).

United States v. *Technicolor, Inc.,* CCH 1950–51 ¶ 62,586 (S.D. Calif. Civil No. 5707–M; Complaint, 1947; Consent Judgments, 1948 and 1950).

United States v. *Telescope Carts, Inc.,* CCH 1952–53 Trade Cases ¶ 67,573 (W.D. Mo. Civil No. 6395; Complaint, 1951; Consent Decree, 1953).

United States v. *Textile Machine Works,* CCH 1950–51 Trade Cases ¶

62,709 (S.D.N.Y. Civil No. 43–671; Complaint, 1947; Consent Judgment, 1950).

United States v. *Timken-Detroit Axle Co.,* CCH 1946–47 Trade Cases ¶ 57,603 (E.D. Mich. Civil No. 5642; Complaint, 1946; Consent Judgment, 1947).

United States v. *Timken Roller Bearing Co.,* 341 U.S. 593 (1951).

United States v. *United Engineering and Foundry Co.,* CCH Current Trade Cases ¶ 67,378 (W.D. Pa. Civil No. 8926; Complaint, 1950; Consent Judgment, 1952).

United States v. *United Shoe Machinery Co.,* 247 U.S. 32 (1918).

United States v. *United Shoe Machinery Corp.,* 110 F. Supp. 295 (D. Mass. 1953).

United States v. *United States Alkali Export Assn.,* 86 F. Supp. 59 (S.D.N.Y. 1949).

United States v. *United States Gypsum Co.,* 333 U.S. 364 (1948).

United States v. *United States Pipe and Foundry Co.,* CCH 1948–49 Trade Cases ¶ 62,285 (D.N.J. Civil No. 10772; Complaint, 1947; Consent Decree, 1948).

United States v. *United States Rubber Co.,* CCH 1953–54 Trade Cases ¶ 67,771 (S.D.N.Y. Civil No. 50–564; Complaint, 1948; Consent Decree, 1954).

United States v. *Universal Button Fastening and Button Co.,* CCH 1948–49 Trade Cases ¶ 62,255 (E.D. Mich. Civil No. 5860; Complaint, 1946; Consent Judgment, 1948).

United States v. *Univis Lens Co.,* 316 U.S. 241 (1942).

United States v. *Vehicular Parking,* 54 F. Supp. 828 (D. Del. 1944); 56 F. Supp. 297 (D. Del. 1944); and 61 F. Supp. 656 (D. Del. 1945); also CCH 1944–45 Trade Cases ¶ 57,266, ¶ 57,277 and ¶ 57,404.

United States v. *Western Precipitation Corp.,* CCH 1946–47 Trade Cases ¶ 57,458 (S.D. Calif. Civil No. 4677; Complaint, 1945; Consent Decree, 1946).

United States v. *Westinghouse Electric and Mfg. Co.,* CCH Current Trade Cases ¶ 67,501 (D.C.N.J. Civil No. 5152; Complaint, 1945; Consent Judgment, 1953).

United States v. *White Cap Co.,* CCH 1948–49 Trade Cases ¶ 62,268 (N.D. Ill. Civil No. 46–C–861; Complaint, 1946; Consent Judgment, 1948).

United States v. *Winslow,* 227 U.S. 202 (1913).

United States Consolidated Seeded Raisin Co. v. *Griffin and Skelley Co.,* 126 Fed. 364 (9th Cir. 1903).

Vitamin Technologists v. *Wisconsin Alumni Research Foundation,* 146 F. 2d 941 (9th Cir. 1945).

Williams v. *Hughes Tool Co.,* 186 F. 2d 278 (10th Cir. 1950); affirming 82 F. Supp. 408 (W.D. Okla. 1949).

Wilson v. *Simpson,* 9 Howard 109 (1850).

Wisconsin Alumni Research Foundation v. *Rene Douglas,* CCH 1945–47 Trade Cases ¶ 57,433 (N.D. Ill. Civil No. 43–C–704; Complaint, 1944; Consent Decree, 1946).

Complaints and Briefs of Department of Justice

United States v. *Associated Patents, Inc.,* (E.D. Mich. Civil No. 10–664; Complaint, 1951).

United States v. *L. D. Caulk Co.* (D. Del. Civil No. 1372; Complaint, 1951).

United States v. *E. I. Du Pont de Nemours and Co.,* CCH 1950–51 Trade Cases ¶ 62,557 and ¶ 62,633 (N.D. Ill. Civil No. 49C–1071; Complaint, 1949).

United States v. *Food Machinery and Chemical Corp.* (N.D. Calif. Civil No. 29308–G; Complaint, 1949).

United States v. *General Railway Signal Co.* (W.D.N.Y. Civil No. 5237; Complaint, 1952).

United States v. *Holophane Co.* (S.D. Ohio Civil No. 2659; Complaint, 1949).

United States v. *International Business Machines Corp.* (S.D.N.Y. Civil No. 72–344; Complaint, 1952).

United States v. *Kelsey-Hayes Wheel Co.* (E.D. Mich. Civil No. 10–655; Complaint, 1951.

United States v. *Kobe* (N.D. Okla. Civil No. 2865; Complaint, 1951).

United States v. *Krasnov* (E.D. Pa. Civil No. 11024; Complaint, 1950).

United States v. *The Logan Co.* (W.D. Pa. Civil No. 96–58; Complaint, 1951).

United States v. *Magnaflux Corp.* (S.D.N.Y. Civil No. 51–C–859; Complaint, 1946).

United States v. *Michigan Tool Co.* (E.D. Mich. Civil No. 12605; Complaint, 1953).

United States v. *Morgan Construction Co.* (D. Mass. Civil No. 52–914; Complaint, 1952).

United States v. *National Malleable and Steel Castings Co.* (N.D. Ohio Civil No. 30281; Complaint, 1953).

United States v. *Ohio Crankshaft Co.* (N.D. Ohio Civil No. 28299; Complaint, 1951).

United States v. *Standard Oil Co.* (*N.J.*) *et al.* (D.C. of D.C. Civil No. 86–27; Complaint, 1953).

United States v. *United Shoe Machinery Corp.*, Brief for the United States, No. 207.

United States v. *Wallace and Tiernan Co.*, CCH 1948–49 Trade Cases ¶ 62,305 and ¶ 62,414 (D.C.R.I. Civil No. 705; Complaint, 1946).

Government Publications

ANNUAL REPORTS

Ann. Rep. Alien Property Custodian.

——— Atomic Energy Commission.

——— Commissioner of Patents.

——— Department of Justice, Office of Alien Property (1951).

——— Federal Trade Commission (1917, 1918, 1919, 1941, 1948); Complaints Nos. 10, 29, 235, 4173, and 5143.

——— National Science Foundation.

HOUSE AND SENATE DOCUMENTS

24 Cong., 1 sess., *Sen. Doc. 338* (1836).

Investigation of Patent Office by President's Commission on Economy and Efficiency, 62 Cong., 3 sess., *House Doc. No. 1110* (1912).

Report of Office of Facts and Figures on Progress of Defense Program, 77 Cong., 2 sess., *Sen. Doc. 157* (1942).

HOUSE AND SENATE HEARINGS

Revision and Codification of the Patent Statutes, Hearings before House Committee on Patents on House Resolution 23417, Parts 1–27, 62 Cong., 2 sess., (Oldfield, 1912).

Revision and Codification of the Patent Statutes, Hearings before House Committee on Patents on House Resolution 15989, Parts 1–10, 63 Cong., 2 sess. (Oldfield, 1914).

United States Patent Office, Hearings before House Committee on Patents on House Resolutions 5011, 5012, and 7010, 66 Cong., 1 sess. (Nolan, 1919).

Pooling of Patents, Hearings before House Committee on Patents on House Resolution 4523, Parts I–IV, 74 Cong.

Investigation of Concentration of Economic Power, Hearings before Temporary National Economic Committee on Public Resolution 113, Parts 2 and 3 (Patents) and Part 25 (Cartels), 75 Cong.

Compulsory Licensing of Patents, Hearings before House Committee on Patents on House Resolutions 9259, 9815, and 1666, 75 Cong., 3 sess.

Patents, Hearings before Senate Committee on Patents on Senate Resolutions 2303 and 2491, Parts 1–9, 77 Cong., 2 sess. (Bone).

Scientific and Technical Mobilization, Hearings before Senate Committee on Military Affairs on Senate Resolutions 107, and 702, Part IV (Patents), 78 Cong., 1 sess.

Elimination of German Resources for War, Hearings before Senate Committee on Military Affairs on Senate Resolution 107 and 146, Parts 1, 2, 3, 4, 10, 79 Cong., 1 and 2 sess.

Study of Monopoly Power, Hearings before House Committee on Judiciary on House Resolutions 137, 156, 6679, and 7905, Serial 14, Parts 1, 2–A, 2–B, 3, 4–A, 4–B, 5, 6–A, 6–B, 81 Cong., 1 and 2 sess.

HOUSE AND SENATE REPORTS

60 Cong., 2 sess., *House Report No. 2145.*

62 Cong., 2 sess., *House Report No. 1161* (Oldfield, 1912).

63 Cong., 2 sess., *House Report No. 1082* (Oldfield, 1914).

Technological Trends and National Policy, 75 Cong., 1 sess., *House Report No. 360.*

Investigation of Concentration of Economic Power, 76 Cong., 1 sess., *House Report No. 95.*

Investigation of the Telephone Industry in the United States, 76 Cong., 1 sess., *House Report No. 340.*

United States versus Economic Concentration and Monopoly, 79 Cong., 2 sess., *House Resolution No. 64.*

SPECIAL REPORTS

American Patent System, National Patent Planning Commission *Report* (1943, 1945).

Dyestuffs for American Textile and Other Industries, U. S. Department of Commerce *Special Agent Series No. 96* (1915).

Edwards, Corwin D. *Economic and Political Aspects of International Cartels.* 78 Cong., 2 sess., *Senate Committee Print Monograph No. 1.*

Farm Machinery Trade Associations, Federal Trade Commission *Report* (1915).

Government-owned Patents and Inventions of Government Employees and Contractors, National Patent Planning Commission *Report* (1944).

Hamilton, Walton. *Patents and Free Enterprise.* 76 Cong., 3 sess., *Senate Committee Print Monograph No. 31.*

Hough, Walter. *Synoptic Series of Objects of the United States National Museum Illustrating History of Inventions,* United States National Museum *Publication,* (1922).

Incandescent Electric Lamps, Tariff Commission *Report No. 133, Second Series* (1937).

Industrial Commission *Report* (1900).

Industrial Research and Development, U. S. Department of Labor and U. S. Department of Defense *Joint Report* (1953).

International Electrical Equipment Cartels, Federal Trade Commission *Report* (1948).

Investigation of Government Patent Practices and Policies: Report and Recommendations of the Attorney General to the President. 3 vols. U. S. Department of Justice, 1946.

National Inventors Council, U. S. Department of Commerce *Information Bulletin* (1951).

Patents, Processes, Techniques, and Inventions, Surplus Property Administration *Report* (1946).

President William H. Taft, *Message to Congress Relative to Patent Laws of the United States,* May 10, 1912.

President Woodrow Wilson, *Annual Message to Congress,* May 20, 1919.

Research and Development Board: Structure and Functions, U. S. Department of Defense *Report* (1952).

Restrictive Business Practices: Analysis of Government Measures Relating to Restrictive Business Practices, United Nations Economic and Social Council *Report.*

Science and Public Policy, President's Scientific Research Board *Report* (1947).

Story of the American Patent System, 1790–1940, U. S. Department of Commerce *Report.*

Tobacco Industry, Corporation Bureau *Report* (1909).

Trust Laws and Unfair Competition, Federal Trade Commission *Report* (1916).

BOOKS AND PAMPHLETS

Arnold, Samuel G. *History of the State of Rhode Island.* 2 vols. Providence, Preston and Rounds, 1899.

Berge, Wendell. *Cartels.* Washington, Public Affairs Press, 1944.

Black, Henry Campbell. *Law Dictionary.* 3rd ed. St. Paul, West Publishing Co., 1933.

Clark, Victor S. *History of Manufacturers in the United States.* 2 vols. Washington, Carnegie Institution of Washington, 1916.

Edwards, Ronald S. *Co-operative Industrial Research.* London, Isaac Pitman and Sons, 1950.

The Federal Antitrust Laws with Summary of Cases, 1890–1951. Washington, Commerce Clearing House, 1952.

Fiske, Bradley A. *Invention, The Master-Key to Progress.* New York, E. P. Dutton and Co., 1921.

Fox, Harold G. *Monopolies and Patents.* Toronto, The University of Toronto Press, 1947.

Gamme, A. A. *Patents of Invention.* London, Longmans, Green and Co., 1946.

Gilfillan, S. C. *The Sociology of Invention.* Chicago, Follett Publishing Co., 1935.

Hearn, William E. *Plutology.* London, Macmillan and Co., 1864.

Hexner, Ervin. *International Cartels.* London, Isaac Pitman and Sons, 1946.

Hobson, John A. *The Evolution of Modern Capitalism.* London, George Allen and Unwin; New York, Charles Scribner's Sons, 1926.

Hubert, P. G., Jr. *Men of Achievement—Inventors.* New York, C. Scribner's Sons, 1893.

Inlow, E. Burke. *The Patent Grant.* Baltimore, The Johns Hopkins Press, 1950.

Jerome, Harry. *Mechanization in Industry.* New York, National Bureau of Economic Research, Inc., 1934.

Kottke, Frank J. *Electrical Technology and the Public Interest.* Washington, American Council on Public Affairs, 1944.

List, Frederich. *The National System of Political Economy.* New York, Longmans, Green and Co., 1928.

Machlup, Fritz. *The Political Economy of Monopoly.* Baltimore, The Johns Hopkins Press, 1952.

Meinhardt, Peter. *Inventions, Patents, and Monopoly.* London, Stevens and Sons, 1950.

Mill, John Stuart. *Principles of Political Economy.* New ed. New York, Longmans, Green and Co., 1936.

Montgomery, R. E. *Industrial Relations in the Chicago Building Trades.* Chicago, 1927; NRC Report, 1937.

Palgrave, R. H. Inglis (ed.). *Dictionary of Political Economy.* 3 vols. London, Macmillan and Co., 1901.

Palmer, Archie M. *University Research and Patent Problems.* Washington, National Research Council, 1949.

Penrose, Edith Tilton. *The Economics of the International Patent System.* Baltimore, The Johns Hopkins Press, 1951.

Perry, John S. *A Defence of the Patent System of the United States.* (U. S. Patent Association *Publications,* Vol. I) Boston, 1875.

Proceedings and Addresses of the Patent Centennial Celebration, 1891. Washington, Gedney and Roberts Co., 1892.

Reese, Jim E. *The Economic Implications of the U. S. Patent System.* Unpublished doctoral thesis, University of Texas, 1940.

Robinson, William C. *The Law of Patents for Useful Inventions.* Boston, Little, Brown and Co., 1890.

Shaler, Nathaniel S. (ed.). *The United States of America.* 2 vols. New York, D. Appleton and Co., 1894.

Smith, Adam. *An Inquiry into the Nature and Cause of the Wealth of Nations.* New York, The Modern Library, 1937.

Summary of Operations and Outline of Services. New York, Manufacturers Aircraft Association, 1951.

Terrell, T. C. and K. E. Shelley. *Patents.* Ninth ed., London, Sweet and Maxwell, 1951.

Thomas, W. I. *Source Book for Social Origins.* Chicago, University of Chicago Press, 1909.

Thompson, Holland. *The Age of Invention.* New Haven, Yale University Press, 1921.

Vaughan, Floyd L. *Economics of Our Patent System.* New York, The Macmillan Co., 1925.

Veblen, Thorstein. *The Instinct of Workmanship.* New York, The Macmillan Co., 1914.

Walker, Francis A. *Discussions in Economics and Statistics.* 2 vols. New York, Henry Holt and Co., 1899.

Wood, Laurence I. *Patents and Antitrust Law.* New York, Commerce Clearing House, 1942.

Periodical Literature and Pamphlets

Currie, Brainerd (ed.) "The Patent System," *Law and Contemporary Problems,* Autumn, 1947, and Spring, 1948.

Cullen, Daniel G. "Patents in Litigation, 1941–45," *Journal of the Patent Office Society,* Vol. XXVIII, No. 12 (Dec., 1946), 903–904.

Kahn, Alfred E. "Fundamental Deficiencies of the American Patent Law," *American Economic Review,* Vol. XXX, No. 3 (Sept., 1940), 475–91.

MacKeigon, I. M. "Notes on 'Patents in Relation to Monopoly,' " *Canadian Journal of Economics and Political Science,* Vol. XII, No. 4 (Nov., 1946), 475.

National Petroleum News, Dec. 8, 1937.

Ooms, Casper W. "The Patent Provisions of the Atomic Energy Act," *University of Chicago Law Review,* Vol. XV, No. 4 (Summer, 1948), 822–38.

Plant, Arnold. "The Economic Theory Concerning Patents for Inventions," *Economica,* Vol. XIV, No. 1 (Feb., 1934), 29–51.

Reik, Richard. "Compulsory Licensing of Patents," *American Economic Review,* Vol. XXXVI, No. 5 (Dec., 1946), 813–32.

Terrill, Robert P. "Cartels and the International Exchange of Technology," *American Economic Review,* Vol. XXXVI, No. 2 (May, 1946), 745–67.

Timberg, Sigmund, "Equitable Relief Under the Sherman Act," *University of Illinois Law Forum,* Winter, 1950, 629–59.

Vaughan, Floyd L. "Important Differences in U. S. and U. K. Patent Systems," *Journal of the Patent Office Society,* Vol. XXXIII, No. 11 (Nov., 1951), 779–99.

Laws, Orders, and Records

Body of Liberties of Massachusetts, 1641, Liberty 9.

Clayton Act of 1914.

Colonial Laws, Massachusetts, 1672.

Constitution of the United States.

Executive Order 9865.

Federal Trade Commission Act of 1914.

Journal, House of Burgesses, Dec. 6, 15, 1764.

Patent Act of 1952. (U.S. Code 35).

Records of the Colony of Massachusetts Bay, Vol. 3, 1646.

Sherman Antitrust Act.

Statute of Monopolies (English).

INDEX